EINAS OSMAN

Essential Medical Mycology

EINAS OSMAN

Essential Medical Mycology

A Comprehensive Laboratory Manual

Noor Publishing

Cover image: www.ingimage.com

Publisher:
Noor Publishing
is a trademark of
Dodo Books Indian Ocean Ltd. and OmniScriptum S.R.L publishing group

120 High Road, East Finchley, London, N2 9ED, United Kingdom
Str. Armeneasca 28/1, office 1, Chisinau MD-2012, Republic of Moldova, Europe
Managing Directors: Ieva Konstantinova, Victoria Ursu
info@omniscriptum.com

Printed at: see last page
ISBN: 978-620-8-87036-2

Essential Medical Mycology: *A Comprehensive Laboratory Manual*

Dr. EINAS OSMAN

Dedication:

This manual is dedicated to my esteemed colleagues in medical laboratories worldwide.

To the tireless laboratory professionals who work diligently behind the scenes, often unrecognized yet essential to patient care. Your precision, dedication, and commitment to excellence form the foundation of modern healthcare.

In mycology laboratories, particularly, where patience meets persistence, where keen eyes distinguish subtle morphological differences, and where scientific rigour yields life-saving diagnoses.

To those who mentor students and share knowledge freely, ensuring the next generation carries forward our standards of excellence.

To my colleagues who face emerging challenges with resilience, adapt to evolving technologies with grace, and maintain an unwavering commitment to quality patient care.

Your work makes a difference every day.

And to my beloved family,

whose love, patience, and constant support have sustained me through every step of this journey. Your encouragement gave me strength, and your belief in me made this work possible.

With profound respect and gratitude,

Einas Osman

May 2025

Biography:

Dr. Einas Osman is an Assistant Professor of Medical Microbiology at the Department of Medical Laboratory Sciences, College of Applied and Health Sciences, A'Sharqiyah University, Oman. With over 12 years of experience in teaching and supervising both undergraduate and postgraduate students in various Sudanese universities, Dr. Osman has developed a deep understanding of academic and research needs in the field of medical microbiology.

Dr. Osman holds a PhD in Molecular Biology from the University of Khartoum, in addition to an MSc in Medical Laboratory Sciences (Microbiology specialization) from the University of Medical Science and Technology, and a BSc in Medical Laboratory Sciences from Sudan University of Science and Technology. Her research primarily focuses on antimicrobial resistance (AMR) in gram-negative bacteria, especially hospital-acquired infections, with an emphasis on their genetic and epidemiological characterization in a global health context. This research has resulted in a range of publications that highlight the diverse and unique epidemiology of bacterial pathogens.

Dr. Osman has also demonstrated significant expertise in the development of postgraduate programs. She served as the Postgraduate Coordinator for four years at the Faculty of Medical Laboratory Sciences, Ibn Sina University, Khartoum, Sudan, where she played a pivotal role in advancing the quality of postgraduate education.

Her academic and research pursuits continue to contribute to a broader understanding of infectious diseases and bacterial genetics, marking her as a distinguished figure in the field.

Preface

This manual is designed as a comprehensive guide for laboratory professionals, students, and researchers engaged in the field of medical mycology. It aims to standardize laboratory procedures, promote best practices, and support the accurate identification of fungal pathogens. Through the clear and practical guidelines presented, users are equipped to enhance the quality, consistency, and reliability of their diagnostic outcomes. Ultimately, this manual contributes to advancing patient care and supporting broader public health efforts in the fight against fungal infections.

Contents

1. Introduction to Medical Mycology Diagnostics

1.1. Purpose of the Manual:

This manual serves as a comprehensive guide for laboratory professionals, students, and researchers in the field of mycology diagnostics. It aims to standardize procedures, ensure accurate identification of fungal pathogens, and promote best practices in mycological testing. By following the guidelines outlined in this manual, users can enhance the quality and reliability of their diagnostic results, ultimately contributing to improved patient care and public health outcomes.

1.2. Global Burden of Fungal Infections:

Fungal infections represent a significant and growing global health challenge. Recent estimates indicate that over 1.7 billion people are affected by fungal infections annually, with approximately 12 million serious infections and 1.5 million deaths per year (Bongomin et al., 2017). The impact varies by region and population:

1.3. Invasive Fungal Infections (IFIs):

- Invasive candidiasis: mortality rates of 30-40%(Kullberg & Arendrup, 2015)

- Invasive aspergillosis: mortality rates reaching 50-60% in high-risk populations(Hoenigl et al., 2012)

- Cryptococcal meningitis: mortality rates up to 70% in resource-limited settings (WHO, 2024)

1.4. Economic Impact:

- Global healthcare costs exceed $7.2 billion annually(Centres for Disease Control and Prevention, 2024)

- Substantial indirect costs due to lost productivity and long-term disability(Richardson et al., 2019)

1.5. Emerging Fungal Pathogens:

The mycological landscape is continuously evolving with new challenges:

1.5.1. Emerging Species:

- *Candida auris*: A multidrug-resistant pathogen showing increased global spread(Jeffery-Smith et al., 2018)

- Novel *Aspergillus spp* with enhanced virulence profiles(Baldin et al., 2021)

- Geographic expansion of endemic mycoses(Seidel et al., 2024)

1.5.2. Environmental Changes:

- Climate change impacts on fungal ecology and distribution(Fisher et al., 2018)

- Emergence of antifungal resistance in environmental isolates(Lockhart et al., 2023)

1.6. High-Risk Populations:

Several populations show increased vulnerability to fungal infections:

1.6.1. Immunocompromised Patients:

- HIV/AIDS patients: 20-25% risk of opportunistic fungal infections(UNAIDS, 2024)

- Organ transplant recipients: up to 30% risk of IFIs(Plantinga et al., 2009)

- Hematologic malignancy patients: 15-40% risk of IFIs during treatment(Cornely et al., 2019)

1.6.2. Healthcare-Associated Risk Factors:

- Central venous catheter use

- Broad-spectrum antibiotic exposure

- Prolonged ICU stay

- Major surgical procedures(Chen et al., 2021)

1.7. Modern Diagnostic Challenges:

Current challenges in mycological diagnostics include:

1.7.1. Technical Limitations:

- Long culture times (3-14 days for definitive identification)

- Variable sensitivity of traditional methods

- Complex specimen requirements

- Difficulty distinguishing colonization from infection(Wickes & Wiederhold, 2018)

1.7.2. Emerging Technologies:

- MALDI-TOF mass spectrometry
- Next-generation sequencing
- Point-of-care molecular testing
- Artificial intelligence applications(Kozel & Wickes, 2014)

1.8. Quality Assurance Requirements:

Quality assurance in medical mycology laboratories is fundamental to providing reliable diagnostic services that directly impact patient care. Modern mycology laboratories must maintain comprehensive quality management systems that align with international standards such as ISO 15189 and meet regulatory requirements. The implementation of robust quality assurance (QA) and quality control (QC) programs ensures consistency, accuracy, and reliability in laboratory results.

1.8.1. Components of Quality Management System

1.8.1.1. Internal Quality Control

1.8.1.1.1. Method Validation

- Validation of new diagnostic methods before implementation
- Verification of commercial systems
- Documentation of analytical performance characteristics
- Establishment of acceptance criteria

1.8.1.1.2. Quality Control Procedures

- Daily controls for staining procedures
- Media quality checks
- Temperature monitoring of incubators and storage
- Equipment calibration and maintenance
- Documentation of all control results

1.8.1.1.3. Personnel Management

- Competency assessment programs
- Continuous education requirements

- Performance monitoring
- Training documentation

1.8.1.2. External Quality Assessment

1.8.1.2.1. Proficiency Testing

- Participation in recognized PT programs
- Analysis of unknown samples
- Performance comparison with peer laboratories
- Implementation of corrective actions when needed

1.8.1.2.2. External Quality Assessment Schemes

- Regular participation in national/international programs
- Analysis of performance trends
- Investigation of discrepant results
- Documentation of corrective actions

1.8.1.3. Documentation and Records

1.8.1.3.1. Standard Operating Procedures

- Detailed written procedures for all processes
- Regular review and updates
- Version control
- Accessibility to all personnel

1.8.1.3.2. Quality Records

- Equipment maintenance logs
- Temperature monitoring records
- Control charts
- Corrective action reports
- Personnel training records

1.8.1.3.3. Quality Indicators

1.8.1.3.3.1. Pre-analytical Phase

- Specimen rejection rates
- Transport time compliance
- Sample adequacy
- Proper identification and labeling

1.8.1.3.3.2. Analytical Phase

- Turn-around time
- Control result compliance
- Equipment performance
- Method performance characteristics

1.8.1.3.3.3. Post-analytical Phase

- Report accuracy
- Result communication
- Clinical correlation
- Customer satisfaction

1.8.1.3.4. Quality Improvement

1.8.1.3.4.1. Continuous Monitoring

- Regular audit programs
- Performance indicator tracking
- Customer feedback analysis
- Non-conformity management

1.8.1.3.4.2. Corrective and Preventive Actions

- Root cause analysis
- Implementation of corrective measures
- Monitoring of effectiveness
- Preventive action planning

1.9. Manual Organization:

This manual provides comprehensive coverage of:

1. Safety precautions and biosafety considerations
2. Specimen collection, transport, and processing
3. Microscopic examination techniques
4. Culture methods and identification procedures
5. Biochemical and molecular testing
6. Antifungal susceptibility testing
7. Quality control and documentation
8. Result interpretation and reporting

2. Safety Precautions in Medical Mycology

2.1. Introduction to Laboratory Safety:

Laboratory safety in medical mycology requires a comprehensive approach due to the potential risks associated with handling fungal pathogens. This section outlines essential safety measures, risk assessment procedures, and emergency protocols based on current international standards and guidelines.

2.2. Risk Assessment:

1. Risk Assessment Framework

Laboratory directors must conduct and document risk assessments for all procedures involving fungal pathogens. The assessment should consider:

- Pathogenicity of the organism

- Route of transmission

- Infectious dose

- Stability in the environment

- Available preventive measures

- Available treatment options(Wold Health Organization, 2024)

Table 1: Risk Classification Matrix

Risk Level	Description	Examples	Required Controls
High	BSL-3 organisms, potential for aerosol transmission	*Coccidioides immitis, Histoplasma capsulatum*	BSL-3 facility, respiratory protection
Moderate	BSL-2 organisms, limited transmission risk	*Aspergillus spp., Candida spp.*	BSL-2 facility, biological safety cabinet
Low	BSL-1 organisms, minimal risk	Non-pathogenic environmental fungi	Standard microbiological practices

2.3. Biosafety Levels and Containment:

2.3.1. BSL-1 Requirements:

- Standard microbiological practices
- Open bench work permitted
- Hand washing facilities
- Decontamination of work surfaces (Centers for Disease Control and Prevention, 2024a)

2.3.2. BSL-2 Requirements:

All BSL-1 requirements plus:

- Biological Safety Cabinet (Class II) for procedures with infection risk
- Restricted laboratory access
- Biohazard warning signs
- Autoclave available
- Eye wash station(American Biological Safety Association International, 2024)

2.3.3. BSL-3 Requirements:

All BSL-2 requirements plus:

- Dedicated ventilation system with HEPA filtration
- Anteroom with changing facilities
- All procedures conducted in BSC
- Respiratory protection program
- Enhanced decontamination protocols(European Biosafety Association, 2024)

2.4. Personal Protective Equipment (PPE):

2.4.1. Basic PPE Requirements

2.4.1.1. Laboratory Coat Specifications:

- Long-sleeved, fluid-resistant
- Fastened front closure

- Knitted or elasticized cuffs

- Changed when visibly soiled

- Removed before leaving laboratory(Occupational Safety and Health Administration, 2024b)

2.4.1.2. Glove Selection

- Nitrile gloves (preferred)

- Double glove for high-risk procedures

- Changed every 30-60 minutes

- Inspected before use

- Removed aseptically

2.4.1.3. Eye Protection:

- Safety glasses with side shields

- Chemical splash goggles

- Face shield when needed

- Impact-resistant lenses

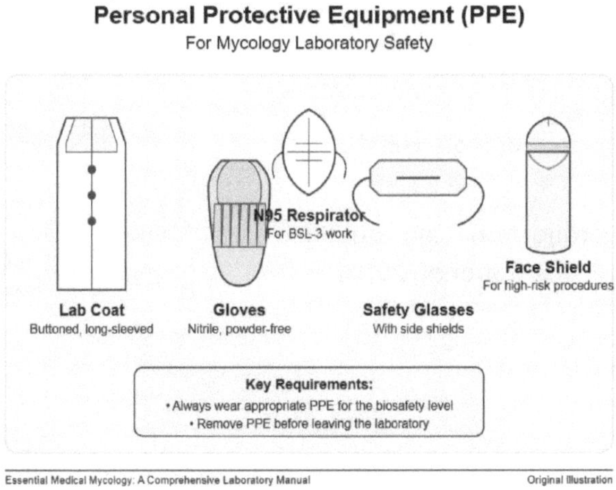

Figure 1: personal protective equipment

2.4.2. Additional PPE for Special Procedures

2.4.2.1. Respiratory Protection:

- N95 respirators for BSL-3 work

- PAPR for high-risk procedures

- Annual fit testing required

- Written respiratory protection program(National Institute for Occupational Safety and Health, 2024)

2.5. Engineering Controls:

2.5.1. Biological Safety Cabinets (BSC)

2.5.1.1. Class II Type A2 BSC:

- Annual certification required

HEPA filtered exhaust

- Inflow velocity: 100 fpm

- Regular monitoring of airflow

- Proper placement away from doors/vents

2.5.1.2. Maintenance Requirements:

- Daily surface decontamination

- Weekly monitoring of airflow indicators

- Monthly inspection of HEPA filter gauges

- Annual certification by qualified technician(NSF International, 2024)

Figure 2: Image of biological safety cabinet
Source 1:https://www.aircleansystems.com/product/protectaire-class-ii-a2-biological-safety-cabinet/

2.5.1.3. Emergency Response Procedures:

1. Spill Response Protocol

Small Spills (<50mL): Alert nearby personnel

2. Don appropriate PPE

3. Cover spill with absorbent material

4. Apply disinfectant concentrically

5. Wait appropriate contact time

6. Dispose as biohazardous waste

2.5.1.4. Large Spills (>50mL):

1. Evacuate the area

2. Post warning signs

3. Notify supervisor and biosafety officer

4. Don enhanced PPE

5. Follow institutional spill response plan(Centers for Disease Control and Prevention, 2024b)

2.6. Exposure Response

2.6.1. Immediate Actions:

1. Remove contaminated PPE

2. Wash exposed area thoroughly

3. Report to supervisor

4. Seek medical evaluation

5. Document incident

2.6.2. Post-Exposure Follow-up:

- Medical surveillance

- Incident investigation

- Protocol review

- Corrective actions implementation(Occupational Safety and Health Administration, 2024a)

2.7. Decontamination and Waste Management:

2.7.1. Surface Decontamination

2.7.1.1. Approved Disinfectants:

- 1:10 sodium hypochlorite
- 70% ethanol or isopropanol
- Quaternary ammonium compounds
- EPA-registered fungicidal agents

2.7.1.2. Contact Times:

- Minimum 10 minutes for routine decontamination
- Extended time for specific organisms
- Documentation of contact time adherence

2. Waste Handling

2.7.1.3. Categories:

- Sharps waste
- Contaminated materials
- Culture media
- Contaminated PPE

2.7.2. Disposal Procedures:

- Use puncture-resistant containers for sharps
- Double-bag biohazardous waste
- Autoclave when required
- Maintain waste logs(Environmental Protection Agency, 2024)

2.7.3. Training Requirements:

2.7.3.1. Initial Training:

- Biosafety principles

- PPE use and maintenance

- Emergency procedures

- Waste handling

- Documentation requirements

2.7.3.2. Ongoing Training:

- Annual refresher courses

- Updates on new procedures

- Incident response drills

- Competency assessments(Clinical and Laboratory Standards Institute., 2024)

2.7.4. Documentation and Records:

1. Required Documentation

- Risk assessments

- Training records

- Incident reports

- Equipment maintenance logs

- Exposure monitoring results

2. Record Retention

- Minimum 3 years for general records

- 30 years for exposure records

- Electronic and physical backup systems(Occupational Safety and Health Administration, 2024c)

3. Specimen Collection and Processing

3.1. Overview

The accuracy of mycological diagnosis depends fundamentally on proper specimen collection, transport, and processing. This section provides comprehensive guidance for handling various clinical specimens in medical mycology.

Proper specimen collection and processing are crucial steps in mycology diagnostics, as they directly impact the accuracy and reliability of test results(Wang et al., 2020). Inadequate or improper specimen collection, transportation, or processing can lead to false-negative results, delayed diagnoses, or inappropriate treatment. This section outlines the types of specimens commonly encountered, collection methods, transportation requirements, and processing techniques.

3.2. Types of specimens

3.2.1. Superficial specimens, skin, hair, and nail scrapings for suspected dermatophyte infections:

These superficial infections, caused by fungi such as *Trichophyton, Microsporum*, and Epidermophyton, typically affect the keratinized tissues of the body. According to recent guidelines of Medical Mycology(American Society for Microbiology, 2024), active lesion borders yield the highest diagnostic sensitivity for dermatophyte infections. Studies have demonstrated that proper sampling technique can increase recovery rates by up to 40% compared to random sampling.

For nail specimens, (Chaya & Pande, 2007)established that subungual debris from the most proximal portion of the infected area provides the highest diagnostic yield. Their multicentre study showed a 65% improvement in culture positivity rates when following standardized collection protocols.

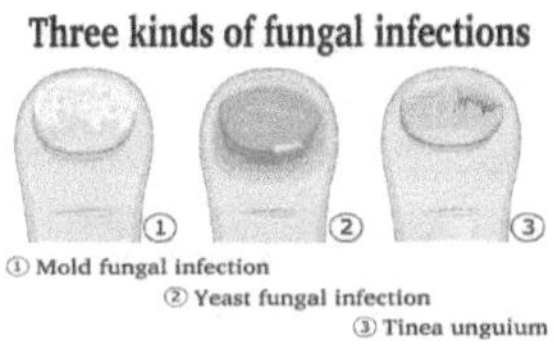

Figure 3: Types of nail fungal infection
Source 2American Academy of Dermatology. (2021). Types of Onychomycosis: Clinical Presentations and Treatment Options. Journal of Dermatological Practice, 14(3), 78-92.:

3.2.2. Respiratory tract infections (sputum or bronchoalveolar lavage (BAL) fluid):

Fungal pathogens like *Aspergillus*, *Cryptococcus*, and *Pneumocystis* can cause pulmonary infections, particularly in immunocompromised patients. The diagnosis of pulmonary fungal infections requires careful attention to specimen quality. Update guidance for processing respiratory tract (ALASTRUEY-IZQUIERDO et al., 2015)demonstrated that early morning sputum specimens, collected after thorough mouth rinsing, provide superior diagnostic yield compared to random collections. For bronchoalveolar lavage (BAL), standardized collection volumes of 40-60 mL per sampled segment optimize fungal recovery.

3.2.3. Blood for suspected systemic infections or fungemia:

Candida, *Cryptococcus*, and other yeast-like fungi can enter the bloodstream and cause life-threatening infections, especially in patients with indwelling catheters, neutropenia, or other risk factors. Current guidelines from the International Society for Human and Animal Mycology (ISHAM) recommend collecting multiple blood culture sets to distinguish true infection from contamination(Chang et al., 2024). A minimum of two sets, each containing 20-40 mL of blood, should be collected from different venipuncture sites. The use of specialized fungal blood culture media has been shown to increase recovery rates by 25-30% for certain pathogens(Qin et al., 2025)

3.2.4. Sterile body fluids (e.g., cerebrospinal fluid (CSF), pleural fluid, peritoneal fluid) for suspected fungal meningitis or other deep-seated infections:

Cryptococcus, *Candida*, and Molds like *Aspergillus* can invade the central nervous system or other sterile body sites. The processing of cerebrospinal fluid (CSF) and other sterile body fluids requires specific protocols to optimize fungal recovery. Recent studies have established minimum volume requirements: 2 mL for cryptococcal antigen testing and at least 5 mL for

fungal culture(Temfack et al., 2021). The European Society for Clinical Microbiology and Infectious Diseases (ESCMID) guidelines emphasize immediate processing for optimal results(Arendrup et al., 2014).

3.2.5. Tissue biopsies from affected organs:

In cases of suspected invasive fungal infections, tissue biopsies may be necessary to confirm the diagnosis and guide treatment. Common sites include the lungs, liver, spleen, and bone.

3.2.6. Swabs from mucosal surfaces (e.g., oral, vaginal):

Candida species are common colonizers of mucosal surfaces and can cause oral thrush or vulvovaginal candidiasis when overgrowth occurs.

3.2.7. Urine for suspected urinary tract infections or disseminated infections:

Fungi like *Candida* can cause urinary tract infections, particularly in patients with urinary catheters or diabetes. The presence of fungi in urine may also indicate disseminated infection.

3.3. Collection methods

The choice of collection method depends on the specimen type and the suspected fungal pathogen. Proper collection techniques are essential to ensure the recovery of the causative organism and minimize contamination. General guidelines for specimen collection include:

- Collect specimens before initiating antifungal therapy, if possible, to maximize the chances of recovery. Antifungal agents can inhibit the growth of fungi in culture and lead to false-negative results. The impact of antifungal therapy on culture yields has been well-documented. A multicentre study by

(Parkins et al., 2007) showed that pre-treatment specimens have 2.5 times higher sensitivity compared to specimens collected after initiating antifungal therapy. When monitoring therapy, standardized collection timing relative to drug administration improves result interpretation.

- Use sterile collection devices and aseptic technique to minimize contamination. This includes using sterile swabs, containers, and instruments, and cleaning the collection site with an appropriate antiseptic solution.

- Collect an adequate amount of specimen to allow for multiple tests or cultures, if necessary. Insufficient specimen quantity can limit the laboratory's ability to perform all required tests and may necessitate recollection.

- For superficial skin, hair, or nail infections, use a scalpel or curette to scrape the affected area, collecting as much material as possible. Samples should be collected from the active border of the lesion, where fungal elements are most likely to be present.

- For sputum samples, instruct patients to collect an early morning, deep-cough specimen after rinsing their mouth with water. This helps to minimize contamination with oral flora and increases the likelihood of obtaining a representative sample from the lower respiratory tract.

- For blood cultures, collect at least two sets from separate venipuncture sites to help differentiate between contamination and true infection. Blood should be collected aseptically and inoculated into fungal blood culture bottles, which contain media designed to support the growth of yeast and Molds.

- For tissue biopsies, obtain specimens from the periphery of the lesion, avoiding necrotic areas if possible. Fungal elements are more likely to be viable in the active edge of the lesion. If the specimen is small, it may be placed in a sterile container with a small amount of sterile saline to prevent desiccation.

3.4. Specimen transportation

Proper transportation of specimens is necessary to maintain the viability of fungal pathogens and prevent contamination. Delays in transportation or exposure to extreme temperatures can impact the recovery of fungi and lead to inaccurate results. A comprehensive analysis by (Dalton et al., 2023) demonstrated optimal recovery rates when specimens are processed within two hours of collection. The American Society for Microbiology (ASM) guidelines provide specific storage conditions for various specimen types when immediate processing isn't possible(Wanger et al., 2017)

Key considerations include:

- Transport specimens to the laboratory promptly after collection, ideally within 2 hours. If processing is delayed, fungal elements may lose viability or become overgrown by commensal bacteria.

- If a delay is unavoidable, store specimens at room temperature or refrigerate at 4°C, depending on the specimen type and suspected pathogen. Most specimens can be held at room temperature for up to 24 hours, but longer delays may require refrigeration. Consult laboratory guidelines for specific storage recommendations.

- Use leak-proof, sterile containers with secure lids to prevent spills and contamination during transport. Specimens should be placed in sealed plastic bags to contain any leakage.

- Label specimens clearly with patient information (e.g., name, date of birth, medical record number), collection date and time, and specimen type. Accurate labelling is essential for proper processing and reporting of results.

- Comply with institutional policies and regulations for packaging and shipping of biological specimens. Follow guidelines for the transportation of dangerous goods when shipping specimens to external laboratories.

3.5. Quality Assurance:

Modern laboratory quality management systems place significant emphasis on monitoring pre-analytical variables due to their substantial impact on laboratory testing accuracy and patient safety. The College of American Pathologists (CAP) checklist underscores the importance of specimen handling and transport(Nielsen, 1997), which are critical components of the pre-analytical phase. This phase is prone to errors, often due to human factors and environmental conditions, which can lead to incorrect test results and patient mismanagement.

3.6. Specimen processing and preparation

Upon receipt in the laboratory, specimens should be processed in a timely manner to optimize recovery of fungal pathogens. Delays in processing can lead to the loss of viability or overgrowth of contaminants. The following steps are generally involved:

- Record the receipt of specimens and verify patient information and labelling. Ensure that the specimen type and collection site match the test requisition.

- Examine the specimen for adequacy and note any visible contamination or deterioration. Reject specimens that are improperly collected, labelled, or transported, as they may yield inaccurate results.

- For skin, hair, or nail specimens, perform a direct microscopic examination using 10-20% potassium hydroxide (KOH) to visualize fungal elements. KOH digests the keratin and cellular debris, allowing the fungal structures to be seen more clearly.

- For sputum or BAL fluid, prepare smears for staining (e.g., Gram stain, calcofluor white) and inoculate appropriate culture media. Gram stain can help to detect the presence of yeast or mold elements and guide the selection of culture media. Calcofluor white is a fluorescent stain that binds to chitin in the fungal cell wall and aids in the visualization of fungal structures.

- For blood, inoculate the specimen into blood culture bottles designed for the recovery of fungi. These bottles contain media that support the growth of yeast and molds and may include antimicrobial agents to inhibit bacterial growth.

- For tissue biopsies, homogenize the specimen and prepare smears for staining and culture. Tissue homogenization helps to release fungal elements from the tissue matrix and increases the likelihood of recovery in culture.

- Inoculate specimens onto appropriate culture media (e.g., Sabouraud dextrose agar, brain heart infusion agar, potato dextrose agar) and incubate at 25-30°C for up to 4 weeks, checking for growth regularly. Different media may be used depending on the suspected pathogen and the specimen type. Incubation at lower temperatures (25-30°C) favors the growth of most pathogenic fungi, while higher temperatures (35-37°C) may be used for the recovery of dimorphic fungi.

By following these guidelines for specimen collection, transportation, and processing, mycology laboratories can ensure the optimal recovery and identification of fungal pathogens. Attention to detail and adherence to established procedures are critical for accurate diagnoses and informed treatment decisions. Ongoing communication between clinicians and laboratory personnel is essential to ensure that specimens are collected, transported, and processed in a manner that maximizes the chances of recovery and minimizes the risk of contamination or inaccurate results.

4. Microscopic Examination in medical mycology

Microscopic examination remains a cornerstone of fungal identification in clinical laboratories. This technique provides rapid, presumptive identification of fungal elements and guides further testing strategies. Recent advances in microscopy techniques and staining methods have enhanced the accuracy and reliability of fungal identification(Guarner & Brandt, 2011). India ink preparation, Giemsa staining, periodic-acid-Schiff staining, Grocott's methenamine silver stain staining, calcofluor mount, and KOH mount are routinely used in diagnostic labs for microscopic examination of clinical specimens.

Mycoses are rapidly diagnosed by observing the presence of fungal pathogens in a clinical sample. This will guide the microbiologist about fungal morphology and helps in making presumptive identification of fungal genera and helps to determine the need for culture and types of culture media to be used(Panizo & Moreno, 2022). Being the easiest, simplest, and cheapest fungal observation technique, it is widely used in diagnostic laboratories.

4.1. Direct Microscopic Examination Techniques:

4.1.1. Potassium Hydroxide (KOH) Mount:

KOH test, also known as KOH mount or KOH preparation, is a rapid test used to visualize the fungal structures in clinical samples using potassium hydroxide (KOH) as a clearing reagent.

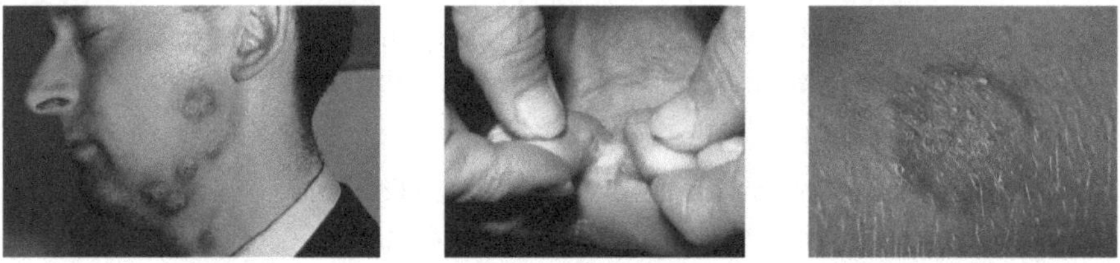

(a) (b) (c)

Figure 4:skin fungal infection

Source 3:https://www.cdc.gov/fungal/diseases/dermatophytes/index.html

4.1.1.1. Objectives of KOH Test:

1. To visualize fungi and their structure in a clinical specimen.
2. To make presumptive identification of dermatophytes.
3. To make a preliminary diagnosis of mycoses.

4.1.1.2. Principle of KOH mount:

The potassium hydroxide dissolves proteinaceous substances including keratin, adhesives that hold keratinized cells together, and other alkali-sensitive tissue materials in clinical specimens(Beg et al., 2023). This digestion results in the breakdown of the cellular components; hence makes the specimen transparent, releases the bound fungal components, and makes the fungal elements clearly visible. The fungal components, however, are alkali resistant so they remain intact. These allow clear visualization of the microscopic morphology of the fungi and fungal elements in the sample.

4.1.1.3. Requirements:

10% to 30% KOH solution is used as a reagent for the KOH mount test.

- 10% KOH: For thin specimens (e.g., skin scrapings)
- 20% KOH: For moderately thick specimens
- 30% KOH: For thick specimens (e.g., nail clippings)

4.1.1.4. Sample/Specimens

All types of clinical specimens can be used for KOH mount. Generally, the specimens include skin scrapings, nail clips, hair, pus, sputum, CSF, tissue (biopsy) sample, urine, mucous membrane swabs, bronchial and alveolar washings, etc.

Specimens determine the concentration of KOH be used. For sputum, pus, CSF, urine, and swabs 10% KOH is effective enough to dissolve tissue debris. For skin scrapings, some swabs with lots of tissue debris, tissue samples, and thick pus, 20% KOH will do the job better and quicker. And for sturdy samples like nail clippings, and hair 30% KOH should be used to get a quick and complete dissolution of debris and keratins.

4.1.1.5. Procedures of KOH test:

4.1.1.5.1. KOH Mount Method

1. In a clean and sterile glass slide, place a drop of KOH. (Concentration of KOH depends on the specimen.)
2. Transfer a small portion of the specimen over the KOH drop and place it on the cover slip on top.
3. Incubate the specimen-KOH mixture at room temperature for 5 to 30 minutes (time varies according to specimen) for clearing the sample and digesting cellular debris.

** Incubate sputum, pus, CSF, urine, and thin pus smear for 5 to 10 minutes.*

** Incubate skin scrapings, thick swabs, thick pus, and tissue samples for about 20 minutes.*

** Incubate hair and nail clippings for 30 minutes.*

Note: Heating the solution over a flame or heating block or incubating at 30^{0}C will accelerate the digestion process. But **DO NOT LET THE KOH BOIL** or **DRY OUT**. To avoid such cases, place the glass slide over moistened filter paper on a petri plate.

4. Examine under a compound light microscope; first at low power (100 X magnification), then shift to high power (400 X magnification). (No need to magnify 1000 times.)
5. Observe the fungal morphology, arrangement of fungal cells/hyphae, morphology and arrangement of fungal spores, and in the case of hair specimen, examine the location of fungi in hairs.

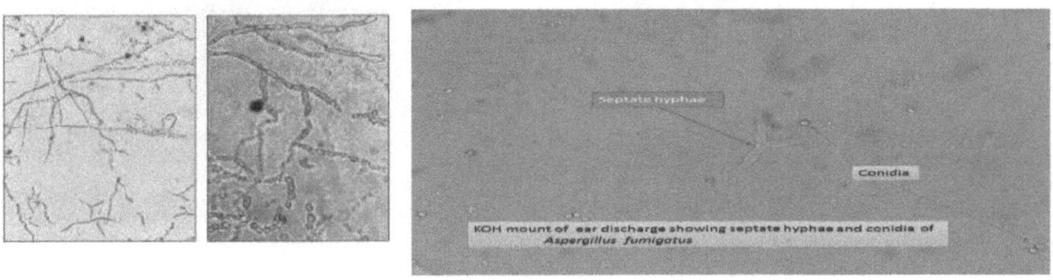

Figure 5:fungal hyphae Structure by using KOH mount technique

Source 4:https://microbeonline.com/koh-mount-principle-procedure-results-uses/

4.1.1.5.2. Modified KOH Mount Methods

KOH mount are modified for better contrast and colouring. Now, following modified KOH mount tests are preferred for better visualization:

4.1.1.5.2.1. KOH-Calcofluor Mount:

In this modified test, all the procedures and requirements are the same, but the reagent is changed from 10 to 30% KOH to 10 to 30% KOH with 0.01% calcofluor-white (a fluorescent blue dye that binds with cellulose and chitin of fungal components) for sample staining.

Preparation of 20% KOH with 0.01% calcofluor white (100 mL)

- Add 20 grams of KOH pellet in 80 mL of distilled water and shake well to dissolve completely.
- In the solution, add 0.1 grams of calcofluor white powder and stir for the complete dissolution of calcofluor crystals.

In this modified test, all the procedures and requirements are the same, but the reagent is changed from 10 to 30% KOH to 20% KOH in 40% DMSO (dimethyl sulfoxide) in distilled water for better clearing of the specimen during visualization.

4.1.1.5.2.2. KOH-DMSO Reagent (100 mL)

- Add 60 grams of DMSO in 90 mL of distilled water to make a 40% DMSO solution.
- In 80 mL of 40% DMSO solution, add 20 grams of KOH pellet and dissolve completely by stirring.

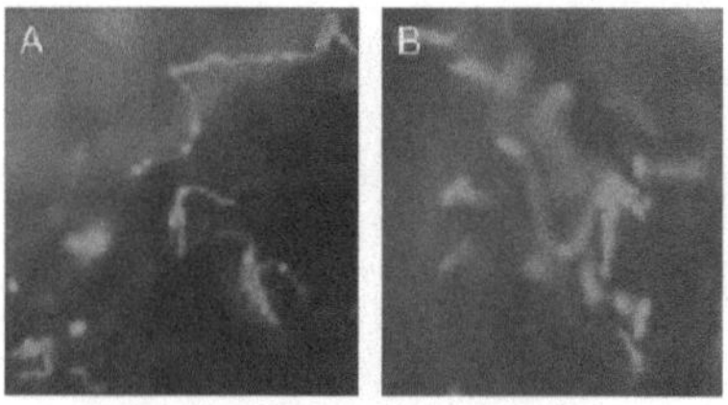

Figure 6:KOH-Calcofluor mount

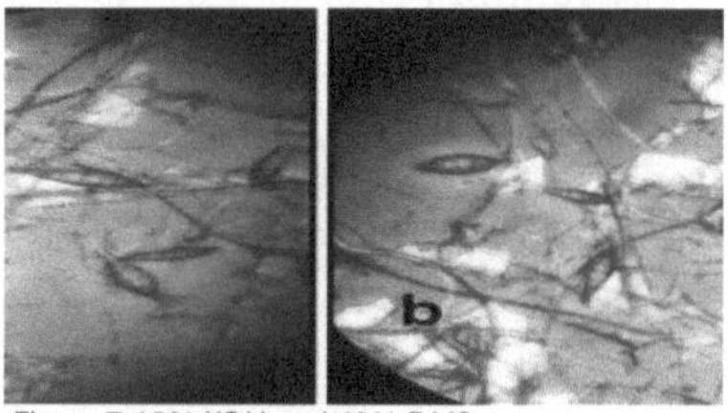

Figure 7:10% KOH and 40% DMS

4.1.2. Staining techniques:

Staining techniques enhance the visualization of fungal structures and aid in their identification. The choice of stain depends on the specimen type, the suspected pathogen, and the desired level of contrast and specificity. The following staining techniques are commonly used in mycology laboratories:

4.1.2.1. Lactophenol Cotton Blue (LPCB):

- Lactophenol Cotton Blue (LPCB) Staining method works on the principle of aiding the identification of the fungal cell walls. The fungal spore cell wall is made up of chitin of which the components of the Lactophenol Cotton Blue solution stains for identification.

- The lactophenol cotton blue solution acts as a mounting solution as well as a staining agent.

- The solution is clear and blue in colour, and it is made up of a combination of three main reagents:

1. Phenol: It acts as a disinfectant by killing any living organisms
2. Lactic acid: To preserve the fungal structures
3. Cotton blue: To stain or give colour to the chitin on the fungal cell wall and other fungal structures

- The stain will give the fungi a blue-coloured appearance of the fungal spores and structures, such as hyphae.

- Mechanism: The phenol and lactic acid act as clearing agents, while the cotton blue stains the fungal cell wall.

4.1.2.1.1. Procedure of Lactophenol Cotton Blue (LPCB) Staining:

1. On a clean microscopic glass slide, add a drop of 70% ethanol
2. Add the fungal specimen to the drop of alcohol using a sterile mounter such as an inoculation loop (from solid medium), depending on the sample of use.
3. Tease the fungal sample of the alcohol using a needle mounter, to ensure the sample mixes well with the alcohol.
4. Using a dropper or pipette, add one or two drops of Lactophenol Cotton Blue Solution before the ethanol dries off.

5. Carefully cover the stain with a clean sterile coverslip without making air bubbles to the stain.
6. Examine the stain microscopically at 40X, to observe for fungal spores and other fungal structures.

4.1.2.1.2. Interpretation:

Fungal structures such as hyphae, conidia, and fruiting bodies appear blue against a clear background. LPCB is particularly useful for the identification of Molds and dermatophytes. For example,

- *Aspergillus niger* stains the hyphae and fruiting structures a delicate blue with a pale blue background.
- *Trichophyton mentagrophytes* also stains the hyphae and fruiting structures a delicate blue with a pale blue background.

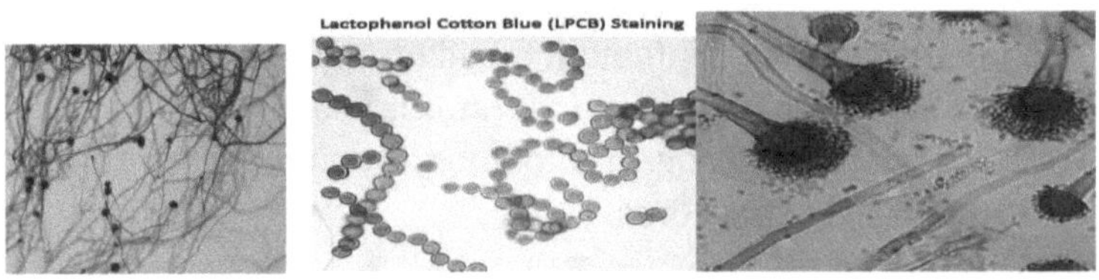

Figure 8:Aspergillus. Flavus LFCB

Source 5:http://www.mycology.adelaide.edu.au/

4.1.2.1.3. Limitation:

- It can only be used as a presumptive identification method of fungi which should be followed up with other diagnostic tools such as biochemical and cultural examination.
- The components of the solution should be used before expiry, including the use of the solution before it expires.
- The solution may disrupt the original morphology of the fungi.
- The stain can only be used to identify mature fungi and its structures and not the young vegetative forms of fungi.
- The stain cannot be stored for a long period of time.

4.1.2.2. Calcofluor White:

4.1.2.2.1. Composition:

Calcofluor white is a fluorescent whitening agent that binds to chitin and cellulose in the fungal cell wall.

4.1.2.2.2. Mechanism:

When exposed to ultraviolet (UV) light, calcofluor white fluoresces bright blue-white, highlighting fungal structures.

4.1.2.2.3. Procedure:

A small amount of clinical specimen (e.g., sputum, tissue) is mixed with a drop of calcofluor white on a microscope slide, covered with a coverslip, and examined under a fluorescence microscope with a UV light source.

4.1.2.2.4. Interpretation:

Fungal elements appear bright blue-white against a dark background. Calcofluor white is highly sensitive for the detection of fungi in clinical specimens but does not provide specific identification.

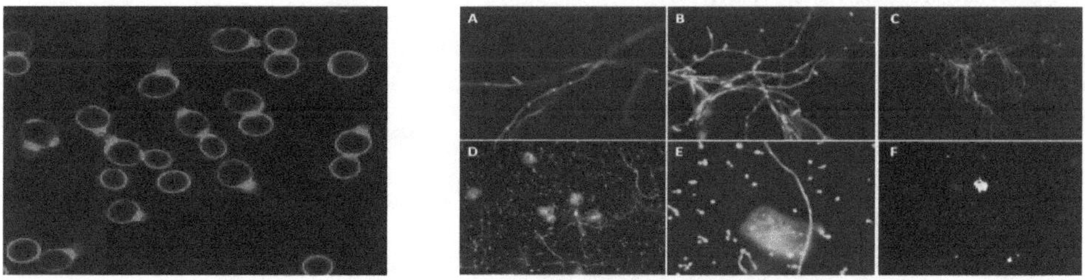

Figure 9:C. albicans and fungal hyphae stained by calcofluor white

Source 6:https://biotium.com/product/calcofluor-white-stain-5-mm-in-water/

4.1.2.3. Gram stain:

4.1.2.3.1. Composition:

Gram stain involves the use of crystal violet, iodine, decolourizer (e.g., acetone or ethanol), and safranin.

4.1.2.3.2. Mechanism:

The staining procedure distinguishes bacteria based on the composition of their cell wall, but it can also be used to visualize yeast cells.

4.1.2.3.3. Procedure:

A thin smear of the clinical specimen is heat-fixed on a microscope slide, stained with crystal violet and iodine, decolorized, and counterstained with safranin. The slide is then examined under the microscope.

4.1.2.3.4. Interpretation:

Yeast cells appear as oval or round structures, staining Gram-positive (purple) or Gram-negative (pink) depending on the species. Gram stain is useful for the rapid detection of yeast in clinical specimens but does not provide definitive identification.

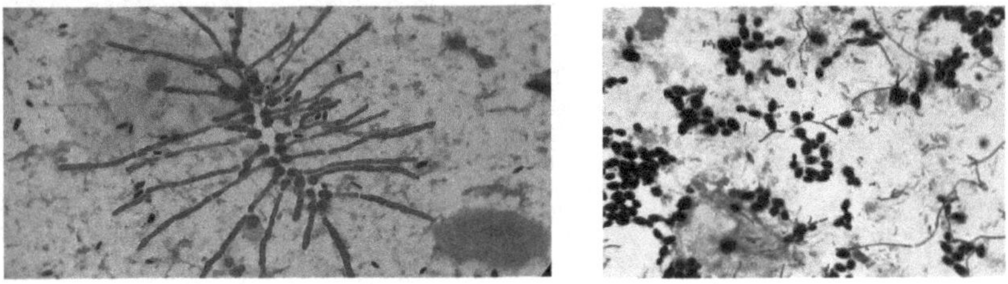

Figure 10:Numerous yeast cells, many budding, along with other bacteria in this sputum sample.

Source 7:American Society for Microbiology. (2023). Clinical Microbiology Procedures Handbook (5th ed.). ASM Press.

4.1.2.4. Giemsa stain:

4.1.2.4.1. Composition:

Giemsa stain is a mixture of methylene blue, eosin, and Azure B.

4.1.2.4.2. Mechanism:

Giemsa stain differentially stains the components of fungal cells, with the cytoplasm appearing blue and the nucleus appearing red or purple.

4.1.2.4.3. Procedure:

A thin smear of the clinical specimen (e.g., CSF, body fluids) is air-dried, fixed with methanol, and stained with Giemsa solution for 15-30 minutes. The slide is then rinsed with water, air-dried, and examined under the microscope.

4.1.2.4.4. Interpretation:

Yeast cells appear as oval or round structures with blue cytoplasm and red or purple nuclei. Giemsa stain is particularly useful for the detection of Cryptococcus in CSF, as the capsule appears as a clear halo surrounding the cell.

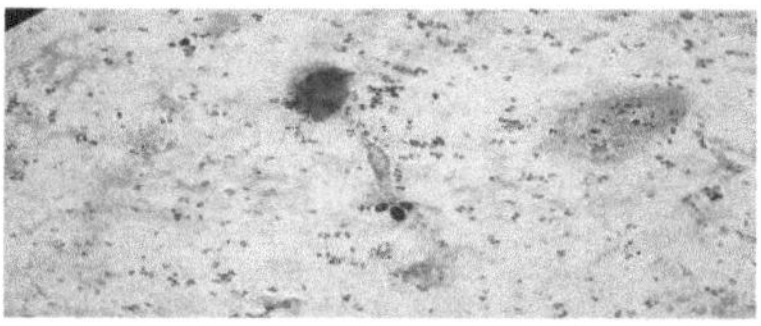

Figure 11:Yeast and hyphae structured seen by Giemsa stain technique

Source 8:https://doi.org/10.1128/CMR.00053-10

4.1.2.5. Gomori Methenamine Silver (GMS) stain:

4.1.2.5.1. Composition:

GMS stain uses methenamine silver nitrate solution and gold chloride.

4.1.2.5.2. principle:

The silver nitrate is reduced to metallic silver, which is deposited on the fungal cell wall.

4.1.2.5.3. Procedure:

Tissue sections are deparaffinized, oxidized with periodic acid, and stained with methenamine silver nitrate solution. The sections are then treated with gold chloride, rinsed, and counterstained with light green or haematoxylin.

4.1.2.5.4. Interpretation:

Fungal elements appear black or brown against a green or blue background. GMS stain is highly sensitive for the detection of fungi in tissue sections and can help to distinguish fungal elements from host tissue.

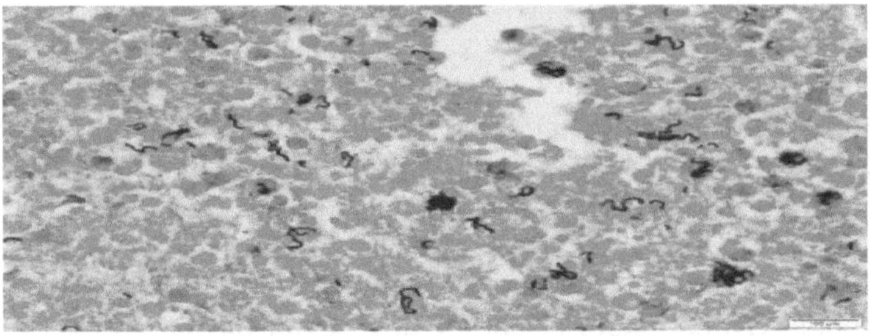

Figure 12:Fungal organisms are highlighted by the Grocott methenamine silver special stain.

Source9:https://www.researchgate.net/publication/344143984_A_Woman_with_a_Lung_Infiltrate_and_Brain_Abscesses_Case_Discussion_from_the_University_of_Louisville_Hospital/figures?lo=1

4.2. Microscopic features of common fungi

The microscopic features of fungi can provide valuable clues to their identity and guide further testing. The following are some of the key microscopic features used in the identification of common fungal pathogens:

4.2.1. Yeast cells:

4.2.1.1. Size and shape:

Yeast cells can vary in size and shape depending on the species. For example, Candida albicans cells are typically oval and 4-6 μm in diameter, while Cryptococcus neoformans cells are round and 5-10 μm in diameter.

4.2.1.2. Budding pattern:

The presence and arrangement of budding cells can be important for identification. For example, C. albicans often forms clusters of budding cells, while C. neoformans typically shows single budding cells with a narrow base.

4.2.1.3. Pseudohyphae:

Some yeast species, such as *C. albicans*, can form pseudohyphae (chains of elongated yeast cells) under certain conditions. The presence and appearance of pseudohyphae can be diagnostic for certain species.

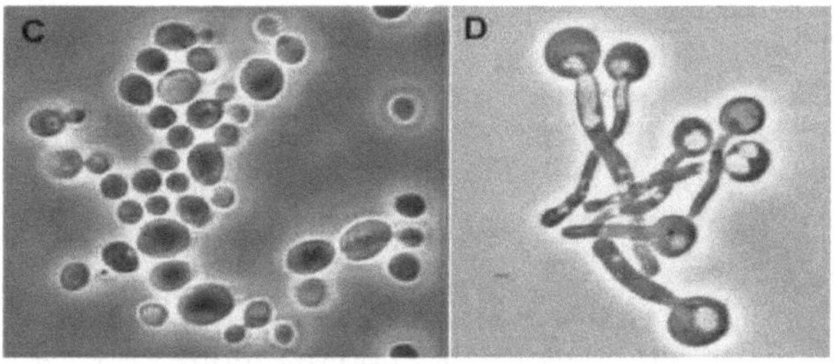

Figure 13:Budding and Pseudo-hyphae Yeast of C. albicans

Source 10:Retrieved from https://safetyculture.com/topics/ppe-safety/

4.2.2. Hyphae:

4.2.2.1. Septation:

Hyphae can be septate (divided by cross-walls) or non-septate (lacking cross-walls). Septate hyphae are characteristic of most Molds, while non-septate hyphae are seen in zygomycetes such as Mucor and Rhizopus.

4.2.2.2. Branching pattern:

The branching pattern of hyphae can vary depending on the species. For example, Aspergillus species typically show acute-angle branching, while Fusarium species show right-angle branching.

4.2.2.3. Hyphal width:

The width of hyphae can be a useful diagnostic feature. For example, the hyphae of dermatophytes are typically narrow (2-3 µm), while the hyphae of Mucor and Rhizopus are typically broad (10-20 µm).

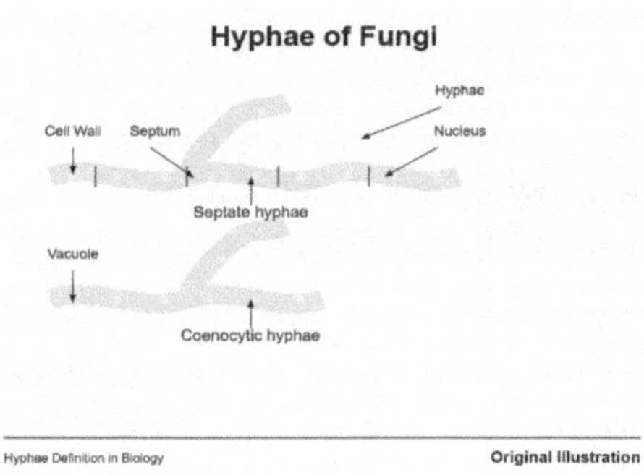

Figure 14:Structure of fungal hyphae

4.2.3. Conidia

4.2.3.1. Shape and size

Conidia can vary in shape and size depending on the species. For example, Aspergillus fumigatus conidia are round and 2-3 μm in diameter, while Fusarium solani conidia are oval and 4-8 μm in length.

4.2.3.2. Arrangement

The arrangement of conidia on the conidiophore can be an important diagnostic feature. For example, Aspergillus species produce chains of conidia from flask-shaped phialides, while Penicillium species produce chains of conidia from brush-like conidiophores.

4.2.3.3. Surface texture

The surface texture of conidia can be smooth or rough, depending on the species. For example, the conidia of A. fumigatus are smooth, while the conidia of A

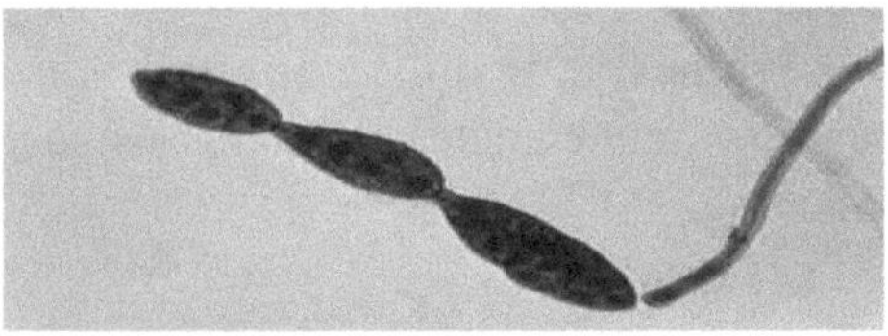

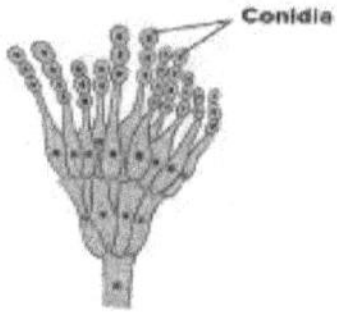

Figure 15:Conidia structure
Source 11:http://phil.cdc.gov/phil_images/20030612/9/PHIL_3963_lores.jpg

4.2.4. Fruiting bodies:

4.2.4.1. Type:

Fungi can produce different types of fruiting bodies, such as cleistothecia (closed fruiting bodies), perithecia (flask-shaped fruiting bodies with an ostiole), or pycnidia (flask-shaped fruiting bodies with conidia).

4.2.4.2. Size and shape:

The size and shape of fruiting bodies can be diagnostic for certain species. For example, the cleistothecia of Trichophyton species are small (100-300 μm) and round, while the perithecia of Chaetomium species are larger (200-500 μm) and flask-shaped.

4.2.4.3. Ascospore arrangement:

The arrangement of ascospores within the fruiting body can be an important diagnostic feature. For example, the ascospores of Trichophyton species are arranged in a linear fashion, while the ascospores of Chaetomium species are arranged in a cluster.

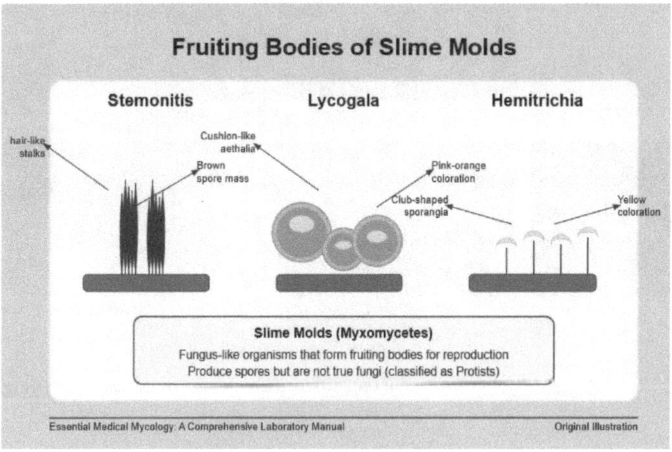

Figure 16:Bizarre fruiting bodies of slime Molds

4.2.5. Capsule:

4.2.5.1. Presence:

The presence of a capsule surrounding the yeast cell is a key diagnostic feature for Cryptococcus species. Other yeast pathogens, such as Candida, do not typically produce a capsule.

4.2.5.2. Thickness:

The thickness of the capsule can vary depending on the species and the clinical presentation. For example, C. neoformans isolates from patients with cryptococcal meningitis often have large, thick capsules, while isolates from patients with pulmonary disease may have smaller, thinner capsules.

4.2.5.3. Staining:

Special stains such as India ink or mucicarmine can be used to visualize the capsule. India ink is a negative stain that outlines the capsule as a clear halo around the cell, while mucicarmine stains the capsule pink or red.

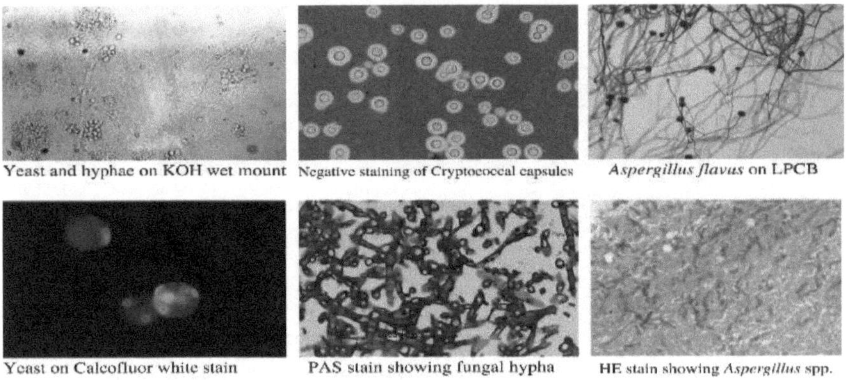

Figure 17:Different capsules structure appearance according to type of stain used.

Source 12:https://microbeonline.com/fungal-staining-methods-and-uses/

5. Culture and Isolation Techniques

Fungal culture is an essential method for isolating and identifying fungal pathogens from clinical specimens. The choice of culture media and incubation conditions depends on the type of fungus suspected and the specimen source.

5.1. Popular Fungi Culture Media

5.1.1. Sabouraud Dextrose Agar (SDA):

A general-purpose medium for the isolation and cultivation of fungi. SDA contains peptone, dextrose, and agar, with a pH of 5.6 to inhibit bacterial growth.

5.1.2. Potato Dextrose Agar (PDA):

A medium rich in carbohydrates, used for the cultivation of fungi that prefer more acidic conditions.

5.1.3. Mycosel Agar:

A selective medium containing cycloheximide, which inhibits the growth of saprophytic fungi and allows the isolation of dermatophytes.

5.1.4. Brain Heart Infusion (BHI) Agar:

A nutrient-rich medium used for the cultivation of dimorphic fungi in their yeast phase.

5.1.5. Chromogenic Media:

Selective and differential media that contain chromogenic substrates, allowing the presumptive identification of certain yeast species based on colony colour.

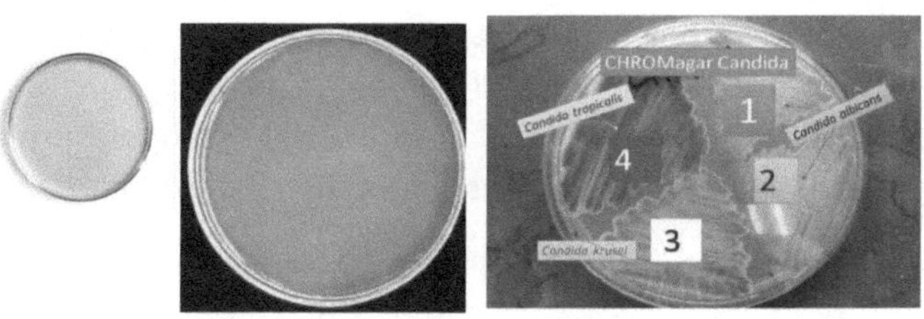

Figure 18:from left to right SDA, Mycosel agar and Candida chromogenic media
Source 13:https://universe84a.com/collection/chromagar/

5.2. Incubation Conditions

5.2.1. Temperature:

Most fungi grow optimally at 25-30°C, while dimorphic fungi require incubation at 35-37°C for the yeast phase and 25-30°C for the mold phase.

5.2.2. Duration:

Incubation times vary depending on the fungal species, ranging from a few days to several weeks.

5.2.3. Atmosphere:

Most fungi grow well in ambient air, but some require increased CO_2 levels or anaerobic conditions.

5.3. Isolation Techniques:

5.3.1. Direct Plating:

Clinical specimens are directly inoculated onto the surface of the culture medium using a sterile loop or swab.

5.3.1.1. Principle:

Direct plating involves inoculating clinical specimens directly onto the surface of the culture medium to isolate fungal pathogens.

5.3.1.2. Procedure:

1. Select appropriate culture media based on the specimen type and suspected fungal pathogen.
2. Using a sterile loop or swab, inoculate the specimen onto the surface of the medium, creating a primary streak.
3. Using a sterile loop, spread the inoculum to obtain isolated colonies by streaking in a zigzag pattern, crossing the primary streak at a 90-degree angle.
4. Incubate the plates at the appropriate temperature and duration based on the suspected fungal pathogen.

5.3.1.3. Advantages:

Simple, rapid, and suitable for specimens with low fungal loads.

5.3.1.4. Limitations:

May not be effective for specimens with high bacterial contamination or low fungal loads.

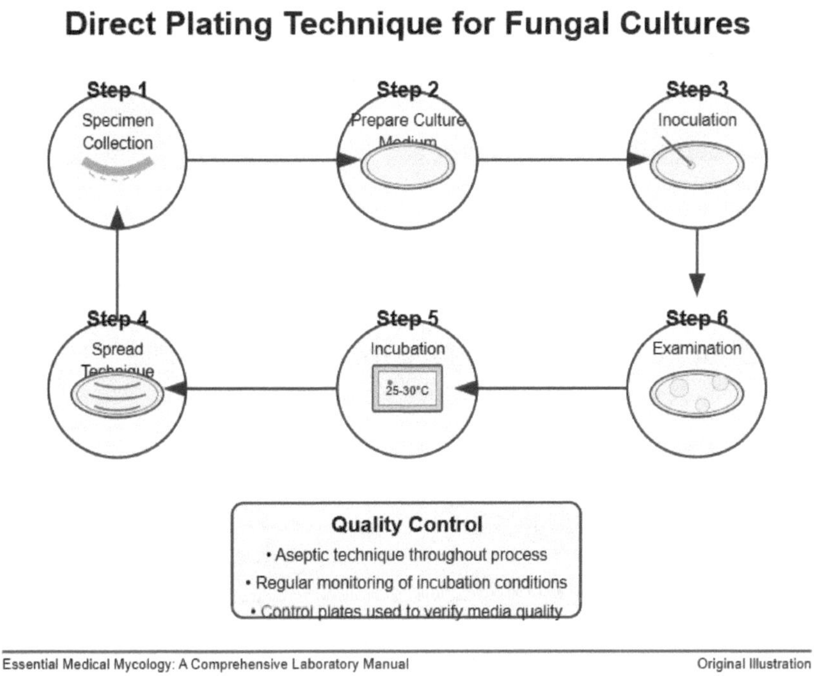

Figure 19:Direct Plating technique

5.3.2. Dilution Plating:

Specimens with high fungal loads are diluted before plating to obtain isolated colonies.

5.3.2.1. Principle:

Dilution plating involves diluting the clinical specimen to reduce the microbial load and obtain isolated fungal colonies.

5.3.2.2. Procedure:

1. Prepare a series of dilutions of the specimen using sterile saline or buffer (e.g., 1:10, 1:100, 1:1000).
2. Inoculate a small volume (e.g., 0.1 mL) of each dilution onto the surface of appropriate culture media.
3. Using a sterile spreader, evenly distribute the inoculum over the surface of the medium.

4. Incubate the plates at the appropriate temperature and duration based on the suspected fungal pathogen.

5.3.2.3. Advantages:

Reduces bacterial contamination and allows the isolation of individual fungal colonies.

5.3.2.4. Limitations:

Requires additional time and materials for dilution preparation.

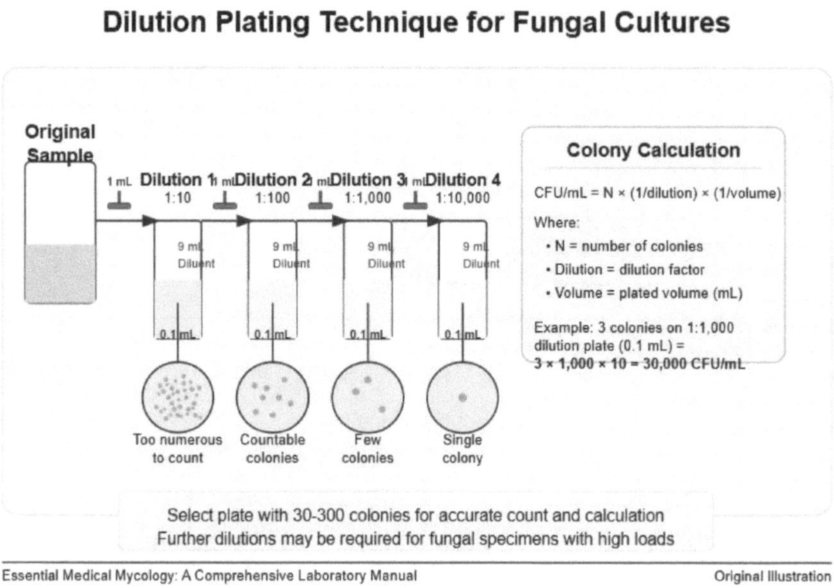

Figure 20:Dilution Plating technique

5.3.3. Pour Plating:

Molten agar is mixed with the specimen and allowed to solidify, enabling the growth of fungi throughout the medium.

5.3.3.1. Principle:

Pour plating involves mixing the clinical specimen with molten agar, allowing the fungal propagules to grow throughout the medium.

5.3.3.2. Procedure:

1. Melt the appropriate culture medium and cool it to approximately 45-50°C.
2. Inoculate a small volume of the specimen (e.g., 0.1-1 mL) into the molten agar.
3. Gently mix the inoculum with the agar by swirling the tube or bottle.
4. Pour the inoculated agar into a sterile Petri dish and allow it to solidify.
5. Incubate the plates at the appropriate temperature and duration based on the suspected fungal pathogen.

5.3.3.3. Advantages:

Allows the growth of fungi throughout the medium, which is useful for specimens with low fungal loads.

5.3.3.4. Limitations:

Colonies may be embedded within the agar, making it difficult to isolate individual colonies.

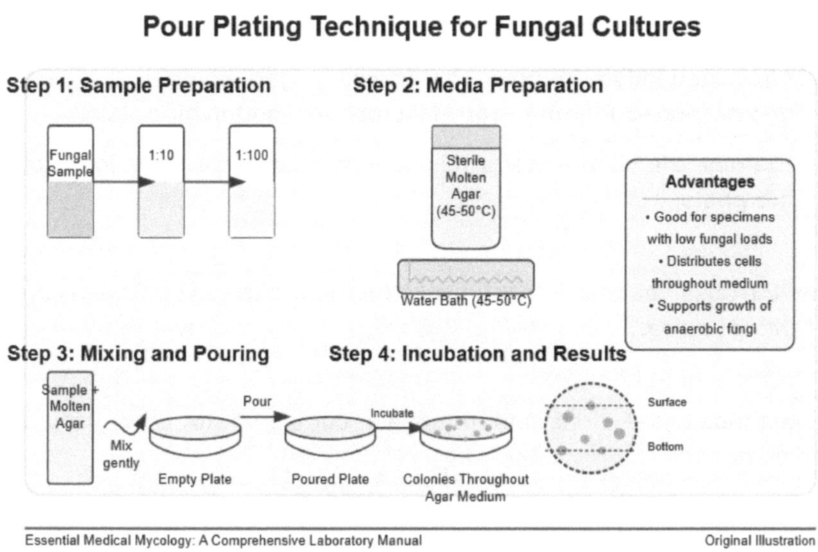

Figure 21:Pour Plating technique

5.3.4. Slide Culture:

A technique used to study the microscopic morphology of fungi by allowing them to grow on a small square of agar placed on a microscope slide.

5.3.4.1. Principle:

Slide culture is a technique used to study the microscopic morphology of fungi by allowing them to grow on a small square of agar placed on a microscope slide.

5.3.4.2. Procedure:

1. Place a small square of appropriate agar medium (e.g., 1 cm x 1 cm) on a sterile microscope slide.
2. Inoculate the four sides of the agar square with the fungal isolate using a sterile loop or needle.
3. Place a sterile coverslip on top of the agar square.
4. Place the slide in a Petri dish lined with moist filter paper to maintain humidity.
5. Incubate the Petri dish at the appropriate temperature and duration based on the fungal isolate.
6. After incubation, remove the coverslip and place it on a new microscope slide with a drop of lactophenol cotton blue stain.
7. Examine the slide under a microscope to observe the microscopic morphology of the fungus.

5.3.4.3. Advantages:

Allows the observation of intact fungal structures, such as conidia, hyphae, and fruiting bodies.

5.3.4.4. Limitations:

Requires additional time and materials compared to other isolation techniques.

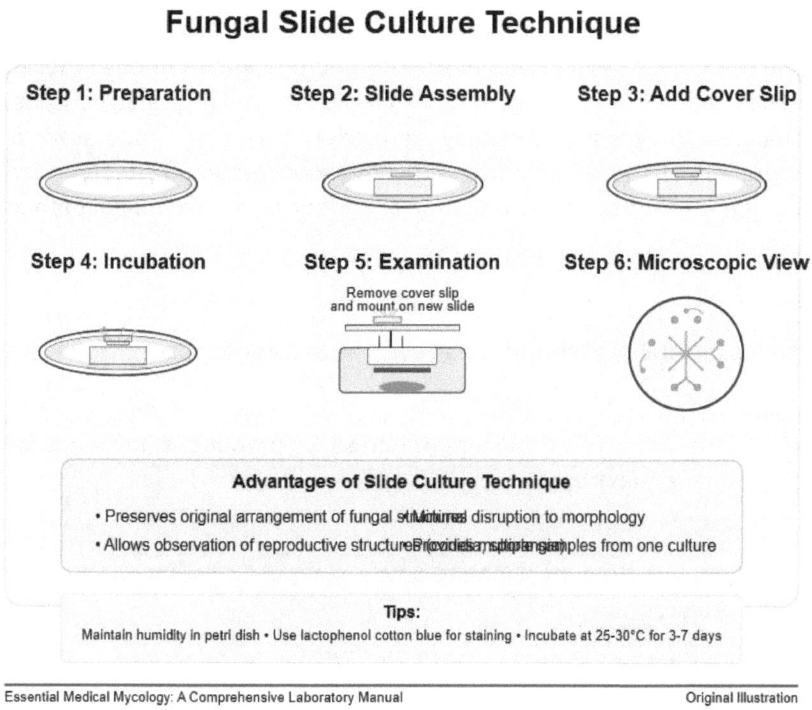

Figure 22:Slide technique

6. Biochemical and Immunological Tests

Biochemical and immunological tests are pivotal in the identification and characterization of fungal pathogens(mousa et al., 2019; Yan et al., 2021), complementing traditional culture and microscopic methods. These tests provide specific insights into fungal metabolism and antigenic composition, enhancing diagnostic accuracy in modern mycology laboratories. The integration of these methods with advanced molecular techniques has significantly improved the detection and identification of fungal pathogens.

6.1. Biochemical Tests:

6.1.1. Urease Test:

Detects the ability of certain fungi, such as Cryptococcus, to hydrolyse urea.

6.1.1.1. Principle:

Detects the ability of certain fungi, such as Cryptococcus, to hydrolyse urea using the enzyme urease.

6.1.1.2. Procedure:

1. Inoculate the fungal isolate onto a slant of Christensen's urea agar or urea broth.
2. Incubate the slant or broth at 25-37°C for 24-72 hours.
3. Observe the slant or broth for a colour change from yellow to pink or red, indicating a positive result.

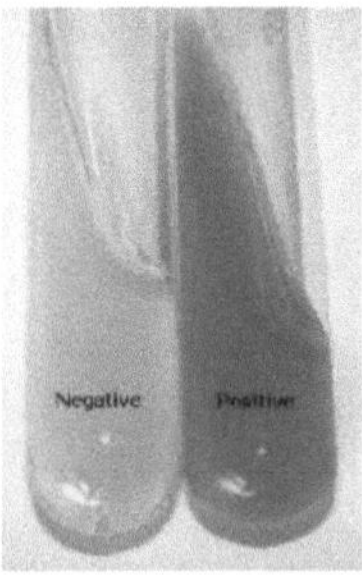

Figure 23: Urease test
Source 14:https://www.medical-labs.net/urease-test-2-2957/#google_vignette

Table 2: Interpretation Guide of urease test

Result	Colour Change	Interpretation	Example Organisms
Positive	Pink-red	Urease Produced	*Cryptococcus spp*
Negative	No change	No urase	*Candida spp*
Invalid	Yellow	Contamination	Repeat test

6.1.1.3. Quality Control

- Positive control: *Cryptococcus neoformans*
- Negative control: *Candida albicans*
- Temperature monitoring during incubation

6.1.2. Nitrate Assimilation Test:

Determines the ability of fungi to utilize nitrate as a sole nitrogen source.

6.1.2.1. Principle:

Determines the ability of fungi to utilize nitrate as a sole nitrogen source.

6.1.2.2. Procedure:

1. Prepare a nitrate assimilation medium containing potassium nitrate as the sole nitrogen source.
2. Inoculate the fungal isolate onto the medium and incubate at 25-30°C for 24-72 hours.
3. Observe the medium for growth, which indicates the ability to assimilate nitrate.

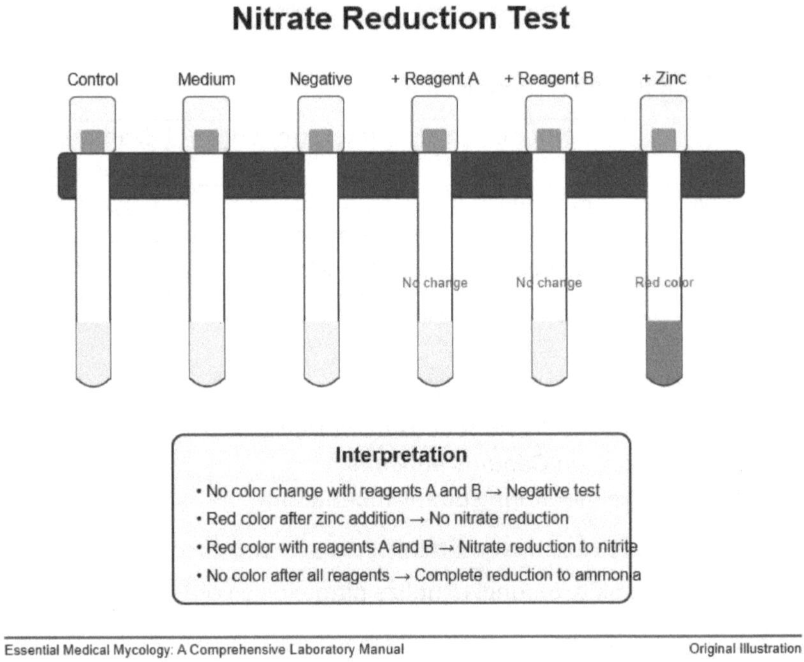

Figure 24:Nitrate Assimilation test

6.1.2.3. Interpretation:

Nitrate assimilation is useful for differentiating certain fungal species, such as distinguishing Candida albicans (nitrate-positive) from Candida dubliniensis (nitrate-negative).

6.1.3. Carbohydrate Assimilation Tests:

Evaluate the ability of fungi to utilize specific carbohydrates as carbon sources.

6.1.3.1. Principle:

Evaluate the ability of fungi to utilize specific carbohydrates as carbon sources.

6.1.3.2. Procedure:

1. Prepare a carbohydrate assimilation medium containing the desired carbohydrate (e.g., glucose, maltose, sucrose, lactose) as the sole carbon source.
2. Inoculate the fungal isolate onto the medium and incubate at 25-37°C for 24-72 hours.
3. Observe the medium for growth, which indicates the ability to assimilate the specific carbohydrate.

Table 3: interpretation results of Common Yeast Identification Patterns

Species	**Glucose**	**Maltose**	**Sucrose**	**Lactose**	**Galactose**
C. albicans	+ve	+ve	+ve	-ve	+ve
C. tropicalis	+ve	+ve	+ve	-ve	+ve
C. glabrata	+ve	-ve	-ve	-ve	-ve
C. krusei	+ve	-ve	-ve	-ve	-ve
C. parapsilosis	+ve	V	+ve	-ve	V

+ve = Positive, **-ve** = Negative, **V** = Variable

Figure 25: Carbohydrate Assimilation Tests
Source15: https://www.researchgate.net/publication/271021722_Erythritol_production_by_Moniliella_megachiliensis_using_non-refined_glycerol_waste_as_carbon_source/figures?lo=1

6.1.4. Germ Tube Test:

A rapid test for the presumptive identification of Candida albicans, based on the formation of germ tubes when incubated in serum at 37°C.

6.1.4.1. Principle:

A rapid test for the presumptive identification of Candida albicans, based on the formation of germ tubes when incubated in serum at 37°C.

6.1.4.2. Procedure:

1. Inoculate a small amount of the yeast colony into 0.5-1 mL of sterile serum (e.g., human, rabbit, or horse serum).
2. Incubate the serum at 37°C for 2-4 hours.
3. Place a drop of the serum on a microscope slide, cover with a coverslip, and examine under a microscope (400x magnification).
4. Look for the presence of germ tubes, which are elongated, cylindrical outgrowths from the yeast cells.

6.1.4.3. Interpretation:

The formation of germ tubes is a presumptive identification of Candida albicans. However, some other Candida species, such as *C. dubliniensis*, can also produce germ tubes.

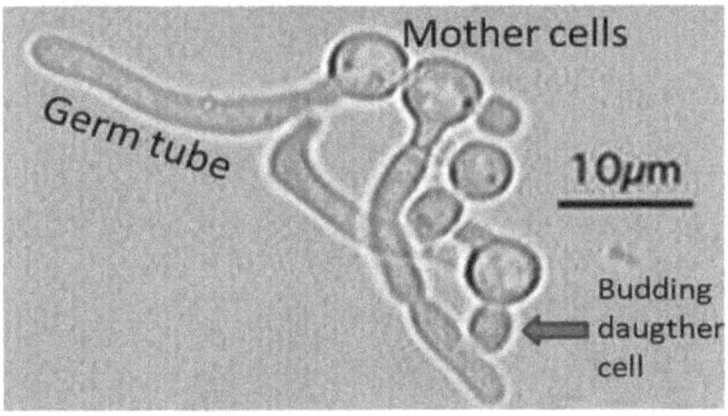

Figure 26:Germ tube test for C. albicans

Source 16:https://commons.wikimedia.org/wiki/File:C_albicans_germ_tubes.jpg

6.1.5. Caffeic Acid Ferric Citrate (CAFC) Test:

6.1.5.1. Principle:

Detects the ability of *Cryptococcus neoformans* to produce melanin when grown on a medium containing caffeic acid and ferric citrate.

6.1.5.2. Procedure:

1. Prepare a CAFC agar plate containing caffeic acid and ferric citrate.
2. Inoculate the fungal isolate onto the CAFC agar and incubate at 25-30°C for 2-5 days.
3. Observe the colony for the development of a brown to black colour, indicating melanin production.

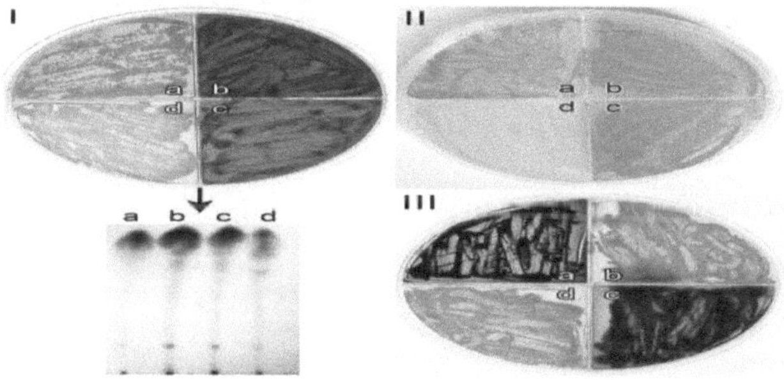

Figure 27:Pigment production by Cryptococcus neoformans in CAFC agar

Source17:https://www.researchgate.net/publication/261742297_Pigment_Production_on_LTryptophan_Medium_by_Cryptococcus_gattii_and_Cryptococcus_neoformans

6.1.5.3. Interpretation:

A positive CAFC test is indicative of *Cryptococcus neoformans*, as other clinically relevant fungi do not produce melanin on this medium

6.1.6. Rapid Trehalose Assimilation Test:

6.1.6.1. Principle:

A rapid test for the presumptive identification of *Candida glabrata*, based on its ability to assimilate trehalose.

6.1.6.2. Procedure:

1. Prepare a trehalose assimilation medium containing trehalose as the sole carbon source.

2. Inoculate the yeast isolate onto the medium and incubate at 37°C for 2-4 hours.
3. Observe the medium for growth, which indicates the ability to assimilate trehalose.

6.1.6.3. Interpretation:

Rapid trehalose assimilation is a characteristic feature of *Candida glabrata*, helping to differentiate it from other clinically relevant *Candida spp*.

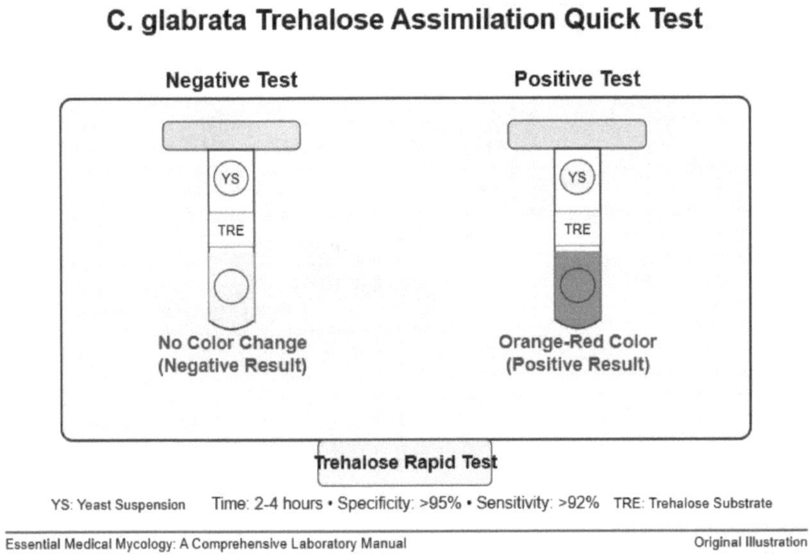

Figure 28:C. glabrata rapid trehalose assimilation quick test

6.2. Immunological Tests:

6.2.1. Galactomannan (GM) Assay:

6.2.1.1. Principle:

Detects galactomannan, a cell wall component of Aspergillus species, in serum or bronchoalveolar lavage fluid.

6.2.1.2. Method:

Enzyme-linked immunosorbent assay (ELISA) or lateral flow device (LFD).

6.2.1.3. Interpretation:

A positive GM assay suggests invasive aspergillosis, but false-positive results can occur in patients receiving certain antibiotics or with other fungal infections.

6.2.2. Cryptococcal Antigen (CrAg) Test:

6.2.2.1. Principle:

Detects the presence of cryptococcal capsular antigen in serum or cerebrospinal fluid.

6.2.2.2. Method:

Latex agglutination (LA), enzyme immunoassay (EIA), or lateral flow assay (LFA).

6.2.2.3. Interpretation:

A positive CrAg test indicates cryptococcal infection, most commonly cryptococcal meningitis.

6.2.3. β-D-Glucan (BDG) Assay:

6.2.3.1. Principle:

Detects β-D-glucan, a cell wall component of various fungi, in serum. It is a broad-spectrum fungal biomarker.

6.2.3.2. Method:

Colorimetric or turbidimetric assay based on the activation of the horseshoe crab coagulation cascade.

6.2.3.3. Interpretation:

A positive BDG assay suggests invasive fungal infection, but it cannot differentiate between specific fungal species. False-positive results can occur in certain clinical settings.

6.2.4. Immunodiffusion (ID) and Complement Fixation (CF) Tests:

6.2.4.1. Principle:

Detect antibodies against specific fungal antigens, such as Histoplasma or Coccidioides.

6.2.4.2. Method:

Immunodiffusion (ID) is based on the formation of precipitin lines between antibodies and antigens in an agar gel. Complement Fixation (CF) measures the consumption of complement in the presence of antigen-antibody complexes.

6.2.4.3. Interpretation:

Positive ID or CF tests indicate exposure to specific fungal pathogens, but they do not necessarily confirm active infection.

6.2.5. Enzyme-Linked Immunosorbent Assay (ELISA):

6.2.5.1. Principle:

Detects antibodies against specific fungal antigens or fungal antigens in clinical specimens.

6.2.5.2. Method:

Indirect ELISA for antibody detection and sandwich ELISA for antigen detection.

6.2.5.3. Interpretation:

Positive ELISA results suggest exposure to or infection with specific fungal pathogens, depending on the antigen or antibody detected.

6.2.6. Immunofluorescence Assay (IFA):

6.2.6.1. Principle:

Detects antibodies against specific fungal antigens using fluorescently labelled antigens or secondary antibodies.

6.2.6.2. Method:

Indirect immunofluorescence assay (IFA) using fungal antigens fixed on microscope slides.

6.2.6.3. Interpretation:

Positive IFA results indicate exposure to specific fungal pathogens, but they should be interpreted in conjunction with clinical and other laboratory findings.

6.2.7. Western Blot (WB) or Immunoblot:

6.2.7.1. Principle:

Detects antibodies against specific fungal antigens separated by electrophoresis and transferred to a membrane.

6.2.7.2. Method:

Fungal antigens are separated by SDS-PAGE, transferred to a membrane, and probed with patient serum. Bound antibodies are detected using enzyme-labelled secondary antibodies.

6.2.7.3. Interpretation:

Western blot allows for the detection of antibodies against specific fungal antigens and can be used for confirmatory testing following a positive screening test.

Table 4: Clinical Scenarios and Recommended Testing Approaches:

Clinical Scenario	Recommended Tests	Alternative Tests	Comments
Suspected IA	Galactomannan	β-D-Glucan	Serial monitoring recommended
Suspected IC	β-D-Glucan	Candida mannan	Consider species identification
CNS infection	Cryptococcal Ag	India ink	CSF preferred sample
Endemic mycosis	Specific Ag test	Antibody detection	Consider geographic exposure

Note: IA = Invasive Aspergillosis, IC = Invasive Candidiasis, CNS = Central Nervous System

Table 5: Quality Control Requirements for immunological tests:

Parameter	Frequency	Acceptance Criteria	Action if out of range
Positive Control	Each run	Expected result ±1 dilution	Repeat test run
Negative Control	Each run	No reaction	Investigate reagents
Temperature	Daily	Within ±1°C of target	Adjust equipment
Reagent QC	Each new lot	Meets specifications	Reject lot

6.3. Additional Considerations in Biochemical and Immunological Testing

The successful implementation of biochemical and immunological tests in medical mycology requires careful consideration of multiple factors beyond the basic testing procedures(Sedik et al., 2024). These additional considerations are crucial for ensuring accurate diagnosis, appropriate test selection, and optimal patient care outcomes. The complexity of fungal infections, combined with the variety of available diagnostic tools, necessitates a systematic approach to test selection and result interpretation.

When selecting and performing diagnostic tests, laboratories must consider various factors including:

1. The specific clinical presentation and suspected pathogen
2. The strengths and limitations of each test
3. The local epidemiology of fungal infections
4. Resource availability and cost considerations
5. Turn-around time requirements
6. The impact of test results on patient manageme

7. Antifungal susceptibility tests:

7.1. Introduction:

The development and evolution of antifungal agents represent a significant advancement in the treatment of fungal infections. As the incidence of invasive fungal infections continues to rise, particularly in immunocompromised populations, understanding the various classes of antifungal agents, their mechanisms of action, and appropriate applications has become increasingly crucial for effective patient care(Wall & Lopez-Ribot, 2020).

The therapeutic arsenal against fungal infections has expanded significantly over the past decades, evolving from relatively simple polyene compounds to include sophisticated targeted therapies. This expansion has been driven by several factors:

1. Rising incidence of invasive fungal infections
2. Emergence of antifungal resistance
3. Growing immunocompromised population
4. Advances in understanding fungal cell biology
5. Development of new drug delivery systems

Recent epidemiological studies indicate that invasive fungal infections affect more than 300 million people annually worldwide, with mortality rates ranging from 20% to 95% depending on the pathogen and host factors(Ikuta et al., 2024). This high disease burden underscores the importance of appropriate antifungal therapy selection and administration.

The evolution of antifungal therapy has been marked by several key developments:

- 1950s: Introduction of polyene antifungals
- 1970s: Development of early azoles
- 1990s: Introduction of triazoles
- 2000s: Development of echinocandins
- 2020s: Emergence of novel drug classes and delivery systems

Current antifungal agents target various aspects of fungal cell structure and metabolism, including:

1. Cell membrane integrity
2. Cell wall synthesis

3. DNA/RNA synthesis
4. Protein synthesis
5. Essential enzyme systems

Understanding these mechanisms of action is crucial for:

- Selecting appropriate therapy
- Predicting potential drug interactions
- Managing adverse effects
- Preventing resistance development
- Optimizing treatment outcomes

7.2. Classes of Antifungal Agents

Antifungal agents are classified into several major groups based on their chemical structure and mode of action, the most widely used and clinically relevant agents in the management of systemic and invasive fungal infections explain in the following table:

Table 6: classes of Antifungal agents

Antifungal Class	Examples	Clinical Applications	Administration Routes
Polyenes	Amphotericin B Nystatin	1. Systemic fungal infections 2. Life-threatening mycoses	1. IV 2. Tropical/ Oral (Nystatin
Azoles	Fluconazole Itraconazole Itraconazole, Voriconazole Posaconazole, Isavuconazole	Broad spectrum antifungal coverage	1. Oral 2. IV (some formulations)
Echinocandins	Caspofungin, Micafungin, Anidulafungin.	1. Invasive candidiasis 2. Aspergillosis 3. Empiric therapy	IV only
Allylamines	Terbinafine	1. Dermatophyte infections 2. Onychomycosis	1. Oral 2. Tropical
Pyrimidine Analogues	Flucytosine (5-FC)	- Cryptococcal infections - Usually combined with other antifungals	Oral

Note: This classification represents the major antifungal classes used in clinical practice. Selection of specific agents should be based on:

1. Type of infection
2. Causative organism
3. Patient factors
4. Local resistance patterns
5. Cost considerations

Beyond the major antifungal classes, several additional agents have established roles in specific clinical scenarios, though with more limited applications. These include:

1. morpholines (e.g., amorolfine), primarily used in topical formulations for dermatophytosis; thiocarbamates (e.g., tolnaftate), effective for superficial fungal infections; and hydroxypyridones (e.g., ciclopirox), which offer broad-spectrum topical activity.
2. Griseofulvin, while historically significant for oral treatment of dermatophyte infections, has been largely superseded by more effective agents: terbinafine (250 mg daily for 6 weeks in tinea corporis), itraconazole (200 mg daily for 1 week per month in onychomycosis), and fluconazole (150-300 mg weekly in tinea capitis). These newer agents offer improved efficacy and shorter treatment durations compared to the traditional 8-12 weeks required with griseofulvin.(Gupta et al., 2018; Kreijkamp-Kaspers et al., 2017)
3. Of particular interest in current drug development are the nikkomycins (e.g., Nikkomycin Z), which target chitin synthase and show promise in clinical trials for systemic fungal infections.

These additional classes complement the major antifungal groups, providing options for specific clinical situations while research continues to explore novel therapeutic approaches(Wiederhold, 2017).

7.3. Mechanism of Action

The mechanisms of action of antifungal agents target different aspects of fungal cell biology:

7.3.1. Polyenes:

Amphotericin B and nystatin bind to ergosterol in the fungal cell membrane, forming pores that lead to membrane disruption, leakage of cellular contents, and cell death.

7.3.2. Azoles:

Fluconazole, itraconazole, voriconazole, posaconazole, and isavuconazole inhibit the fungal cytochrome P450 enzyme lanosterol 14α-demethylase, which is essential for the synthesis of ergosterol. This leads to the accumulation of toxic sterol intermediates and cell membrane dysfunction.

7.3.3. Echinocandins:

Caspofungin, micafungin, and anidulafungin inhibit the synthesis of β-1,3-glucan, a critical component of the fungal cell wall. This results in osmotic instability and cell lysis.

7.3.4. Allylamines:

Terbinafine inhibits the enzyme squalene epoxidase, which is involved in the early stages of ergosterol biosynthesis. This leads to the accumulation of squalene and cell membrane disruption.

7.3.5. Pyrimidine Analogues:

Flucytosine is converted into 5-fluorouracil within the fungal cell, which interferes with DNA and RNA synthesis, leading to the inhibition of fungal growth and replication.

7.4. Indications and Dosing:

The choice of antifungal agent and dosing regimen depends on the type and severity of the fungal infection, the patient's immune status, and other factors such as drug interactions and adverse effects.

7.4.1. Amphotericin B:

Used for severe systemic fungal infections, including invasive aspergillosis, cryptococcal meningitis, and mucormycosis. Dosing varies based on the formulation (deoxycholate or lipid complex) and indication.

7.4.2. Fluconazole:

Used for the treatment of candidiasis, cryptococcosis, and prophylaxis in immunocompromised patients. Dosing ranges from 200-800 mg/day depending on the indication.

7.4.3. Itraconazole:

Used for the treatment of aspergillosis, blastomycosis, histoplasmosis, and onychomycosis. Dosing ranges from 200-400 mg/day.

7.4.4. Voriconazole:

Used for the treatment of invasive aspergillosis, candidiasis, and other serious fungal infections. Dosing is weight-based, with a loading dose followed by maintenance therapy.

7.4.5. Posaconazole:

Used for the prophylaxis of invasive fungal infections in high-risk patients and the treatment of refractory invasive fungal infections. Dosing varies based on the formulation (oral suspension or delayed-release tablet).

7.4.6. Echinocandins:

Used for the treatment of invasive candidiasis, aspergillosis, and other serious fungal infections. Dosing varies based on the specific agent and indication.

7.4.7. Terbinafine:

Used for the treatment of dermatophyte infections, including onychomycosis and tinea pedis. Dosing is typically 250 mg/day for oral therapy.

7.4.8. Flucytosine:

Used in combination with amphotericin B for the treatment of cryptococcal meningitis and other serious fungal infections. Dosing is based on renal function, with a target peak serum concentration of 30-80 μg/mL.

7.5. Adverse Effects and Drug Interactions

Antifungal agents can cause various adverse effects and interact with other medications:

7.5.1. Amphotericin B:

Nephrotoxicity, electrolyte disturbances (hypokalemia, hypomagnesemia), infusion-related reactions (fever, chills, rigors), and anemia. Drug interactions include increased toxicity with other nephrotoxic agents.

7.5.2. Azoles:

Gastrointestinal disturbances (nausea, vomiting, diarrhea), hepatotoxicity, and QT prolongation. Azoles are metabolized by and inhibit cytochrome P450 enzymes, leading to numerous drug interactions.

7.5.3. Echinocandins:

Generally well-tolerated, with occasional gastrointestinal disturbances, headache, and infusion-related reactions. Minimal drug interactions.

7.5.4. Terbinafine:

Gastrointestinal disturbances, headache, and rare cases of hepatotoxicity and skin reactions. Few significant drug interactions.

7.5.5. Flucytosine:

Bone marrow suppression (leukopenia, thrombocytopenia), gastrointestinal disturbances, and hepatotoxicity. Dose adjustment is necessary for renal impairment.

It is essential for healthcare providers to monitor patients for adverse effects and manage drug interactions when using antifungal agents. Dose adjustments may be necessary based on the patient's renal and hepatic function, as well as concomitant medications.

7.6. Principles of Antifungal Susceptibility Testing:

Antifungal susceptibility testing is performed to determine the sensitivity or resistance of a fungal isolate to specific antifungal agents. The main objectives of antifungal susceptibility testing are:

1. To guide the selection of appropriate antifungal therapy
2. To monitor the emergence of antifungal resistance
3. To assess the epidemiology of antifungal resistance

Antifungal susceptibility testing methods should be standardized, reproducible, and clinically relevant. The Clinical and Laboratory Standards Institute (CLSI) and the European Committee on Antimicrobial Susceptibility Testing (EUCAST) provide guidelines for performing and interpreting antifungal susceptibility tests.

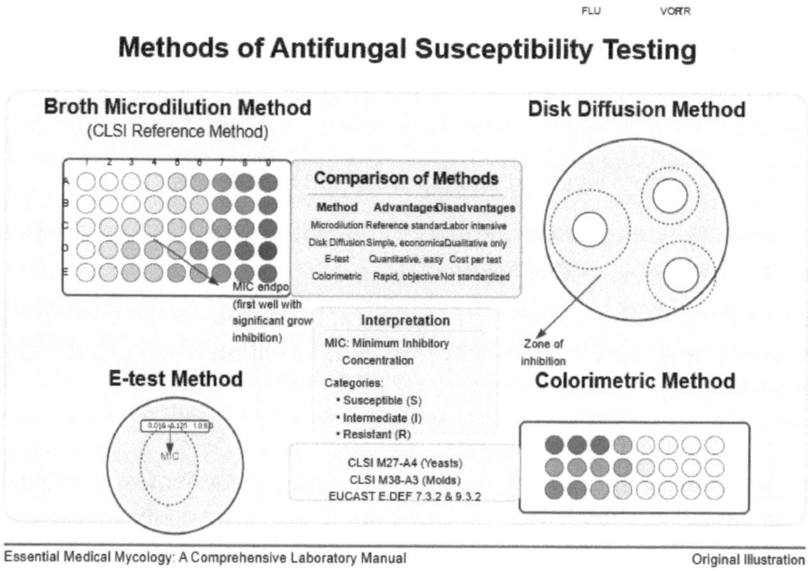

Figure 29:different methods of fungal susceptibility test

7.7. Standardized Testing Methods:

7.7.1. Broth Dilution Methods:

Broth dilution methods are considered the gold standard for antifungal susceptibility testing. There are two main types of broth dilution methods:

7.7.1.1. Broth Macro-dilution:

This traditional method involves testing in tubes containing 1 mL or more of broth medium. While considered a reference method, it has largely been superseded by microdilution techniques due to their greater efficiency and reduced reagent requirements.

- Performed in test tubes with a volume of 1 mL or more

- Antifungal agents are serially diluted in liquid growth medium (RPMI 1640 with MOPS buffer)

- Fungal inoculum is added to each tube, and the tubes are incubated

- The lowest concentration of the antifungal agent that inhibits visible fungal growth is determined as the Minimum Inhibitory Concentration (MIC)

7.7.1.2. Broth Micro-dilution

- Performed in 96-well microtiter plates with a volume of 200 µL or less

- Antifungal agents are serially diluted in liquid growth medium across the wells

- Fungal inoculum is added to each well, and the plate is incubated

- The MIC is determined as the lowest concentration of the antifungal agent that inhibits visible fungal growth

Broth microdilution is the most used method due to its convenience, reproducibility, and the ability to test multiple antifungal agents simultaneously.

7.7.2. Disk Diffusion Method

The disk diffusion method is a simple and cost-effective method for antifungal susceptibility testing. However, it is not as widely used as broth dilution methods due to limitations in reproducibility and the lack of established interpretive criteria for many fungal species and antifungal agents.

7.7.2.1. Procedure:

- A standardized fungal inoculum is evenly spread on the surface of an agar plate

- Paper disks impregnated with specific concentrations of antifungal agents are placed on the inoculated agar surface

- The plate is incubated, and the diameter of the zone of inhibition around each disk is measured

7.7.2.2. Interpretation:

- The zone of inhibition is compared to established interpretive criteria, when available, to categorize the fungal isolate as susceptible, intermediate, or resistant to the antifungal agent

7.7.2.3. Interpretation of Results:

The interpretation of antifungal susceptibility testing results depends on the method used and the availability of established interpretive criteria.

7.7.2.3.1. Broth Dilution Methods:

- The MIC is the primary endpoint

- CLSI and EUCAST provide species-specific clinical breakpoints (CBPs) and epidemiological cutoff values (ECVs) for interpreting MICs

- CBPs categorize isolates as susceptible, intermediate, or resistant based on the likelihood of clinical success

- ECVs distinguish wild-type isolates from those with acquired or mutational resistance

7.7.2.3.2. Disk Diffusion Method:

- The zone of inhibition diameter is the primary endpoint

- Interpretive criteria, when available, categorize isolates as susceptible, intermediate, or resistant

- Interpretive criteria are less well-established for the disk diffusion method compared to broth dilution methods

It is essential to consider the limitations of antifungal susceptibility testing, such as the lack of established interpretive criteria for some fungal species and antifungal agents, and the potential discordance between in vitro susceptibility results and clinical outcomes. Antifungal susceptibility testing results should be interpreted in conjunction with clinical findings, patient factors, and pharmacokinetic/pharmacodynamic considerations.

8. Commercial identification systems:

8.1. VITEK 2 system:

This automated identification system uses a combination of biochemical tests and a proprietary algorithm to identify yeast and Mold species. It includes a database of over 300 species and can provide identification results in as little as 4 hours.

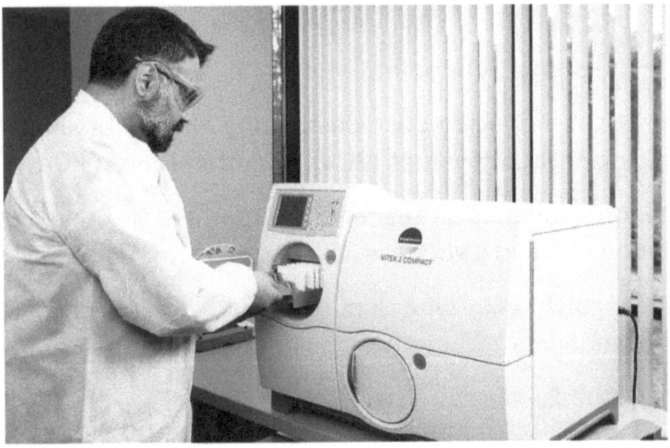

Figure 30:VITEK 2 system

Source 18:https://www.bmxclinicaldiagnostics.com/post/mantenimiento-preventivo

8.2. API 20C AUX system:

This manual identification system uses a battery of 20 biochemical tests to identify yeast species. It includes a database of over 40 species and can provide identification results in 48-72 hours.

Figure 31:analytical profile index API

Source 19:https://www.mdgsb.com.my/products/biomerieux-api--the-global-reference-for-identification

8.3. MALDI Biotype system:

This mass spectrometry-based identification system uses a combination of protein profiles and a proprietary algorithm to identify yeast and Mold species. It includes a database of over 2,000 species and can provide identification results in minutes.

Figure 32:MALDI system

Source 20:https://www.beckmancoulter.com/products/microbiology/bruker-maldi-biotyper-system

By combining the results of microscopic examination with other diagnostic techniques such as culture and biochemical testing, and by using appropriate identification keys and resources, mycology laboratories can accurately identify fungal pathogens and guide appropriate treatment.

Ongoing training and proficiency testing are essential to ensure that laboratory personnel are skilled in the microscopic identification of fungi and can recognize atypical or unusual organisms.

Collaboration with clinical teams and other laboratory disciplines is also important to ensure that the results of microscopic examination are interpreted in the context of the patient's clinical presentation and other diagnostic findings.

References:

- ALASTRUEY-IZQUIERDO, A., MELHEM, M. S. C., BONFIETTI, L. X., & RODRIGUEZ-TUDELA, J. L. (2015). Susceptibility test for fungi: Clinical and laboratorial correlations in medical mycology. *Revista Do Instituto de Medicina Tropical de São Paulo, 57*(suppl 19), 57–64. https://doi.org/10.1590/S0036-46652015000700011

- American Biological Safety Association International. (2024). *Biosafety guidelines for medical mycology laboratories.*

- American Society for Microbiology. (2024). *Guidelines for specimen management in medical mycology laboratories.* DC: ASM Press.

- ARENDRUP, M. C., BOEKHOUT, T., AKOVA, M., MEIS, J. F., CORNELY, O. A., & LORTHOLARY, O. (2014). ESCMID and ECMM joint clinical guidelines for the diagnosis and management of rare invasive yeast infections. *Clinical Microbiology and Infection, 20*, 76–98. https://doi.org/10.1111/1469-0691.12360

- BALDIN, C., KÜHBACHER, A., MERSCHAK, P., SASTRÉ-VELÁSQUEZ, L. E., ABT, B., DIETL, A.-M., HAAS, H., & GSALLER, F. (2021). Inducible selectable marker genes to improve *Aspergillus fumigatus* genetic manipulation. *Journal of Fungi, 7*(7), 506. https://doi.org/10.3390/jof7070506

- BEG, S. S., KALSOOM, F., HUSSAIN, S., & GUL, J. (2023). KOH mount: A prompt and efficient diagnostic tool for mycotic keratitis. *Annals of Punjab Medical College, 17*(2), 258–261. https://doi.org/10.29054/APMC/2023.1323

- BONGOMIN, F., GAGO, S., OLADELE, R., & DENNING, D. (2017). Global and multi-national prevalence of fungal diseases—Estimate precision. *Journal of Fungi, 3*(4), 57. https://doi.org/10.3390/jof3040057

- Centers for Disease Control and Prevention. (2024a). *Biosafety in microbiological and biomedical laboratories* (7th ed.). CDC.

- Centers for Disease Control and Prevention. (2024b). *Emergency response guide for biological laboratories.*

- Centers for Disease Control and Prevention. (2024c). *Fungal disease burden report.* Atlanta: CDC.

- CHANG, C. C., HARRISON, T. S., BICANIC, T. A., CHAYAKULKEEREE, M., SORRELL, T. C., WARRIS, A., HAGEN, F., SPEC, A., OLADELE, R., GOVENDER, N. P., CHEN, S. C., MODY, C. H., GROLL, A. H., CHEN, Y.-C.,

LIONAKIS, M. S., ALANIO, A., CASTAÑEDA, E., LIZARAZO, J., VIDAL, J. E., ... PERFECT, J. R. (2024). Global guideline for the diagnosis and management of cryptococcosis: An initiative of the ECMM and ISHAM in cooperation with the ASM. *The Lancet Infectious Diseases, 24*(8), e495–e512. https://doi.org/10.1016/S1473-3099(23)00731-4

- CHAYA, A., & PANDE, S. (2007). Methods of specimen collection for diagnosis of superficial and subcutaneous fungal infections. *Indian Journal of Dermatology, Venereology and Leprology, 73*(3), 202. https://doi.org/10.4103/0378-6323.32753
- Clinical and Laboratory Standards Institute. (2024). *Laboratory safety guidelines*. CLSI Document (GP17-A4 ed.).
- FISHER, M. C., HAWKINS, N. J., SANGLARD, D., & GURR, S. J. (2018). Worldwide emergence of resistance to antifungal drugs challenges human health and food security. *Science, 360*(6390), 739–742. https://doi.org/10.1126/science.aap7999
- GUARNER, J., & BRANDT, M. E. (2011). Histopathologic diagnosis of fungal infections in the 21st century. *Clinical Microbiology Reviews, 24*(2), 247–280. https://doi.org/10.1128/CMR.00053-10
- National Institute for Occupational Safety and Health. (2024). *Respiratory protection in healthcare settings*.
- Occupational Safety and Health Administration. (2024a). *Exposure control plan guidelines*.
- Occupational Safety and Health Administration. (2024b). *Laboratory safety guidance*.
- World Health Organization. (2024). *Global report on fungal infections*.
- WICKES, B. L., & WIEDERHOLD, N. P. (2018). Molecular diagnostics in medical mycology. *Nature Communications,* 9(1), 5135. https://doi.org/10.1038/s41467-018-07556-5
- Wold Health Organization. (2024). *Laboratory biosafety manual* (5th ed.). WHO Press.

Table of Figures and Sources:

Core Textbooks:

a. Larone, D. H., & Larone, D. H. (1987). *Medically important fungi: a guide to identification* (Vol. 196, p. 203). New York: Elsevier.
   This textbook provides detailed descriptions and illustrations of the microscopic and macroscopic features of over 150 medically important fungi. It includes identification keys, differential diagnosis tables, and case studies to aid in the identification of unknown isolates.
b. Dismukes, W. E., Pappas, P. G., & Sobel, J. D. (Eds.). (2003). Clinical mycology.
   This comprehensive textbook covers the epidemiology, pathogenesis, diagnosis, and treatment of fungal infections. It includes chapters on laboratory diagnosis, with detailed descriptions and images of the microscopic features of common fungal pathogens.
c. Seifert, K. A., & Gams, W. (2011). The genera of Hyphomycetes–2011 update. *Persoonia-Molecular Phylogeny and Evolution of Fungi*, *27*(1), 119-129.
   This textbook provides a comprehensive guide to the identification of hyphomycetes (molds) based on their microscopic features. It includes detailed descriptions, illustrations, and identification keys for over 150 genera of hyphomycetes.

Online databases:

a. Mycology Online (http://www.mycology.adelaide.edu.au/): This online resource provides a wealth of information on the identification and classification of fungi. It includes an image database with over 2,000 images of fungal structures, as well as identification keys, glossaries, and teaching materials.
b. Atlas of Clinical Fungi (http://www.clinicalfungi.org/): This online atlas provides detailed descriptions and images of the microscopic and macroscopic features of over 500 species of medically important fungi. It includes identification keys, differential diagnosis tables, and case studies to aid in the identification of unknown isolates.
c. MycoBank (http://www.mycobank.org/): This online database provides nomenclatural and taxonomic information on over 500,000 fungal species. It includes descriptions, illustrations, and references for each species, as well as links to other online resources.